T0321720

Thermal physics

MACMILLAN PHYSICAL SCIENCE

Series advisers

Physics titles: Dr R L Havill, *University of Sheffield*
                          Dr A K Walton, *University of Sheffield*

Chemistry titles: Dr D M Adams, *University of Leicester*
                          Dr M Green, *University of York*

Titles in the series

Group Theory for Chemists, *G Davidson*
Thermal Physics, *M T Sprackling*
Low Temperature Physics, *A Kent*

MACMILLAN PHYSICAL SCIENCE SERIES

Thermal physics

Michael Sprackling
Dept of Physics, King's College, London

This book is volume one of the
Macmillan Physical Science Series

This edition published 1991 in the
United States and Canada

First published 1991 by
Macmillan Education Ltd,
Houndmills, Basingstoke, Hampshire RG21 2XS and London

Library of Congress Cataloging-in-Publication Data
Sprackling, M. T. (Michael Thomas)
Thermal physics / Michael Sprackling.
p.   cm.
Includes bibliographical references and index.
ISBN 978-0-88318-919-1. — ISBN 978-0-88318-920-7 (pbk.)
1. Heat.   2. Thermodynamics.   I. Title.
QC254.2.S77   1991
536—dc20   91–17550
                                                      CIP

Contents

Preface

Thermal physics is a well-established subject. It developed from the classical thermodynamics that grew out of a study of the behaviour of heat engines in the early part of the nineteenth century, and the basic theory was virtually complete by the end of that century. However, despite its practical origins and the large number of books written about it, many students find thermal physics a 'theoretical' and difficult subject. One reason for this is that many of the concepts are rather subtle, so that the student, while able to follow the mathematics involved, is unable to appreciate the physics. Consequently, the subject appears as a vast collection of equations that lacks a structure and an outlook, and students are unable to apply the principles of thermal physics to new situations.

In this book I have tried to provide an account of the elements of thermal physics, suitable for a first introduction to the subject, that will appeal to undergraduates in physics and engineering. The treatment is based on my experience of teaching thermal physics to undergraduates in the University of London for a number of years. The subject is treated from an experimental standpoint, basic concepts are treated in some detail and, it is hoped, a clear picture is presented of the essential features and characteristics of thermal physics without making the subject appear forbidding.

SI units are used throughout this book. Some worked examples are included, as well as a number of problems for the reader to attempt. These problems form an essential part of the book and should be worked through carefully. Solutions are provided for most of the problems.

It is impossible to write a book on thermal physics without being influenced by the work of others. I am indebted to many authors and I have found the following books particularly valuable: *Heat and Thermodynamics* by M. W. Zemansky, *Elements of Classical Thermodynamics* by A. B. Pippard, *Thermal Physics* by P. C. Riedi, *Equilibrium Thermodynamics* by C. J. Adkins and *Thermal Physics* by C. P. B. Finn. I am also indebted to F. C. Frank for several stimulating conversations.

London, 1990   M.T.S.

Nomenclature

A	area
a	thermal diffusivity
B	availability
B_a	applied magnetic flux density
C	heat capacity
C_p	heat capacity at constant pressure
C_V	heat capacity at constant volume
$C_{p,m}$	molar heat capacity at constant pressure
c	specific heat capacity, molecular speed, speed of light in vacuum
c_p	specific heat capacity at constant pressure
c_s	speed of sound in a gas
d	molecular separation
E	electric field strength, e.m.f., Young's modulus, irradiance within a cavity
E_p	potential energy
E_k	kinetic energy
F	Helmholtz function, Faraday's constant, load, force, form factor
G	Gibbs function
G_m	molar Gibbs function
g	specific Gibbs function, acceleration of free fall
H	enthalpy
H_a	applied magnetic field strength
H_m	molar enthalpy
h	specific enthalpy, heat transfer coefficient (surface emissivity), height, Planck's constant
I	electric current
J	Massieu function
K	bulk modulus
k	Boltzmann's constant, thermal conductivity
L	length
l	mean free path, specific latent heat
M	magnetisation

M_m	mass of one mole
m	magnetic moment, molecular mass, mass
N_A	Avogadro's constant
n	number density of molecules
p	pressure
$\hat{p}$	reduced pressure
Q	heat, quantity of heat, electric charge
q	infinitesimal quantity of heat
R	resistance, radius, radiant emittance, gas constant
r	resistance, radius, compression factor
S	entropy
S_m	molar entropy
s	specific entropy, perimeter
T	thermodynamic temperature
$\hat{T}$	reduced temperature
T^*	magnetic temperature
t	Celsius temperature, time
U	internal energy
U_m	molar internal energy
$\bar{U}$	total heat transfer coefficient
u	specific internal energy, radiant energy per unit volume
u,v,w	velocity components
V	volume
$\hat{V}$	reduced volume
V_m	molar volume
v	specific volume, fluid speed
W	work, quantity of work
W_F	flow work
w	infinitesimal quantity of work
X	generalised force
x	generalised displacement, distance, thickness
Z	stored charge, compressibility factor
α	absorptance (absorption factor), angle of rotation
β	cubic expansivity (isobaric expansivity)
γ	ratio of the principal heat capacities, electrical conductivity
η	thermal efficiency, viscosity
θ	empirical temperature, angle
μ_0	permeability of a vacuum
μ	isenthalpic Joule–Thomson coefficient
ρ	density
σ	Stefan–Boltzmann constant, specific surface free energy, molecular diameter
κ	compressibility
λ	linear expansivity, wavelength
ω	angular velocity

φ	magnetic flux, radiant flux, angle, isothermal Joule–Thomson coefficient
χ_m	magnetic susceptibility
τ	torque, efficiency temperature function

Dimensionless groups

(Nu)	Nusselt number
(Pr)	Prandtl number
(Re)	Reynolds number
(Gr)	Grashof number

Subscripts

l	liquid phase
g	gas phase
v	vaporisation
s	sublimation
f	fusion

1

What is thermal physics?

The term *thermal physics* is applied to the study of the properties of large assemblies of atoms — that is, to the study of macroscopic bodies — and to the interactions between such bodies, when there are differences of temperature between the bodies or when their properties are temperature-dependent.

A study of the properties of a large assembly of atoms can be made from both a large-scale, or macroscopic, viewpoint and an atomic, or microscopic, viewpoint. Thus, ultimately, thermal physics should provide a systematic description of the bulk properties of matter at all temperatures and an explanation of these properties in terms of the constituent atoms and the forces between them.

Very broadly, thermal physics deals with the transfer of energy to, from and between macroscopic bodies. Before the early years of the nineteenth century the law of conservation of energy dealt only with kinetic and potential energy and was valid only when the forces involved were the so-called conservative forces, such as the force of gravity. The law failed when viscous or frictional forces were present. Attempts to incorporate these dissipative forces into a conservation law led to the development of thermal physics, which, as a consequence, has incorporated in it many concepts taken from mechanics.

Macroscopic thermal physics is not concerned with the behaviour of the individual atoms in a body but with the bulk properties of the body, and may be considered to have started with the theory of heat engines published by Sadi Carnot in 1824. The development of Carnot's approach, principally by Clausius and William Thomson (later Lord Kelvin), resulted in the subject being cast effectively in its present form by the end of the nineteenth century. This branch of thermal physics is often called thermodynamics, a name which reflects its engineering origins. Discussion of the behaviour of bodies in terms of their macroscopic properties is often limited to conditions of equilibrium, when the subject is called equilibrium thermodynamics or classical thermodynamics.

In contrast, the microscopic approach seeks to explain the behaviour of matter in terms of the behaviour of the atoms of the body concerned. This approach requires a model of the assembly of atoms that is being considered and of the forces between them, and all such models are necessarily simplifications to some degree. Microscopic models of matter explaining the behaviour of gases date back to the eighteenth century, but the first major developments in this area occurred in the middle and second half of the nineteenth century, with the work of Herapath,, Waterston, Maxwell, Clausius and Boltzmann.

This book is largely concerned with an elementary treatment of the macroscopic thermodynamics of bodies that are in a state of equilibrium. There is one chapter on the kinetic theory of gases which shows how, on the basis of a suitable model, the macroscopic behaviour of dilute gases may be explained in terms of the behaviour of the constituent molecules.

The primary aim of classical thermodynamics is to describe the behaviour of bodies in equilibrium in such a way that, when the equilibrium conditions are altered, the behaviour can be predicted quantitatively — that is, it is concerned with equilibrium conditions and not with the details of the process. However, in real situations the rate at which energy is transferred from one body to another is often important, and the last chapter deals with this topic from a macroscopic viewpoint.

In the remainder of this chapter the outlook and capabilities of classical thermodynamics will be considered briefly.

1.1 Classical thermodynamics

Classical thermodynamics deals with the interaction of bodies from a macroscopic viewpoint. The strength of thermodynamics lies in the fact that only a few well-defined and directly measurable quantities are needed to specify the condition of a body. For example, the condition of a fixed mass of gas is completely specified when the pressure and volume are known. Values of other appropriate quantities may be deduced by thermodynamical arguments, so that relatively few properties of the body need independent explanation. In fact, one of the major objectives of thermodynamics is to derive interrelations between quantities that are independently measurable. Thermodynamical arguments can also be used to deduce values for quantities under conditions where measurement is difficult and to check the consistency of the measurements made.

What often appears to be a limitation of macroscopic thermodynamics is that the internal structure of the body concerned does not play a part in the description. It follows that, although the solutions to problems are always correct, they are not very informative, in the sense that nothing is learned directly about the detailed interactions of the atoms of which the body is composed. None the less, macroscopic thermodynamics provides a check on the models adopted in any microscopic treatment, since the predictions of atomic models must, when scaled up to bodies of macroscopic size, agree with the results of macroscopic thermodynamics.

A consequence of the small number of variables used in the macroscopic description of the behaviour of bodies is that thermodynamics is only able to give an exact description of equilibrium conditions and, therefore, can only provide a limiting description of real processes: the most powerful results in thermodynamics take the form of inequalities, becoming equalities only under idealised conditions.

Nevertheless, Maxwell [1] described thermodynamics as 'a science with secure foundations, clear definitions and distinct boundaries'. It is hoped that this book will bring out these features of the subject. Thermodynamic analysis is valuable in a wide range of applications that include steam and internal combustion engines, electricity generation, the liquefaction of gases, crystal growth, refrigerators and heat pumps, meteorology and weather forecasting, chemical reactions, astrophysics and the "heat balance" of living organisms. Some of these topics are discussed in this book.

Note

1 J. C. Maxwell, *Scientific Papers*, Vol. 2, p. 662 (New York: Dover, 1965).

2

Systems and processes

This chapter introduces, in rather general terms, the classical thermodynamic description of the behaviour of matter. The description is a macroscopic one — that is, in terms of large-scale or gross properties — and, although this has the advantage of not needing any assumptions about the structure of matter [1], it will be shown that the description is then restricted to time-independent situations. The chapter starts by defining a number of terms used in the thermodynamic description.

2.1 Systems

The usual approach to any topic in science is to examine the behaviour of an appropriate region of the universe that can be effectively separated, sometimes only in the imagination, from the rest. The term *system* is used to describe the region of space that is under consideration and the matter that it contains.

A system is of macroscopic dimensions and may be of any degree of complexity, although the majority of systems considered in this book are so-called simple systems — that is, they are macroscopically homogeneous and isotropic, and are influenced to a negligible effect by surface effects, by shear stresses and by external fields such as the gravitational field.

The remainder of the universe outside the system is known as the *surroundings* of the system, although, in practice, the term is reserved for those parts of the environment that have a direct effect upon the behaviour of the system.

The interface between the system and its surroundings is known as the *system boundary* and, for a clear specification of the system, it is vital that the boundary be well-defined (see Figure 2.1a). A system may then be defined as an identifiable collection of matter enclosed by a surface known as the boundary.

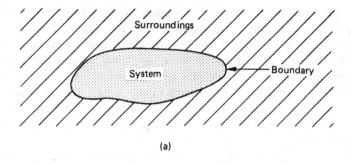

(a)

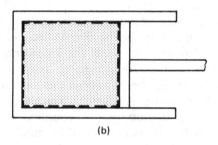

(b)

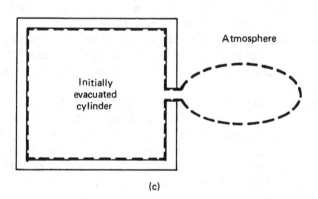

(c)

Figure 2.1 *Illustrating the terms 'system', 'boundary' and 'surroundings'*

The system boundary is always notional, although it may coincide with a real, physical interface. For example, if the system is the gas contained in a cylinder by means of a well-fitting piston, the boundary (shown dashed in Figure 2.1b) coincides with the inner surfaces of the cylinder and piston;

the cylinder wall and the piston themselves are part of the surroundings. In contrast, if a mass of air flows from the atmosphere into an initially evacuated cylinder by way of a valve, and the mass of air that finally enters the cylinder is taken as the system, at some intermediate stage the boundary outside the cylinder may be taken as an imaginary membrane enclosing the air, and shown dashed in Figure 2.1(c). These examples show also that a boundary may change in shape and that the volume that a boundary encloses may alter. It is also possible for the boundary to be permeable to matter.

Classical thermodynamics gives a description of the behaviour of a system and what takes place at the system boundary in terms of the bulk physical properties of the system. This is a macroscopic description: the quantities used in the description apply to the system as a whole. For example, the behaviour of a system consisting of a fixed mass of gas, as in Figure 2.1(b), may be described in terms of the pressure and volume of the gas. The quantities used to describe the macroscopic behaviour of a system are often called the *thermodynamic variables* or *properties*, but in this book they will be referred to as the *thermodynamic coordinates*.

In a first description of the behaviour of a particular system, the quantities used are those that are directly measurable, and such quantities are known as *primitive* thermodynamic coordinates. The number of coordinates needed to describe completely the behaviour of a system must be found by experiment. Essentially, this means that one person must give such a macroscopic description of the system that a second person is able to construct from it a system which is identical with the one described, to the agreed accuracy, and which is in the same condition as the first when the coordinates chosen have been given the same values.

The condition of a system at any instant, identified by a particular set of values of its coordinates, is called the *state* of the system [2]. For this reason thermodynamic coordinates are sometimes called *state variables* or *state functions*.

An important implication of the classical thermodynamic description of the state of a system is that the coordinates refer to the system as a whole. Therefore, a coordinate must be well-defined and, if appropriate, have a uniform value throughout the system. The system will then be in a state in which the values of the coordinates are independent of time, since gradients in the values of the coordinates would lead to changes in the state of the system. For example, when a system consisting of a mass of gas, contained in a cylinder by means of a well-fitting piston (as in Figure 2.2a), is made to expand by a rapid movement of the piston, finite pressure gradients are set up in the gas and the behaviour cannot be described by a single value of the pressure. Furthermore, while the piston is moving, the actual space occupied by the gas may not correspond to the geometrical volume of the cylinder as indicated by the position of the piston, shown schematically in Figure 2.2(b). When the piston stops, the pressure

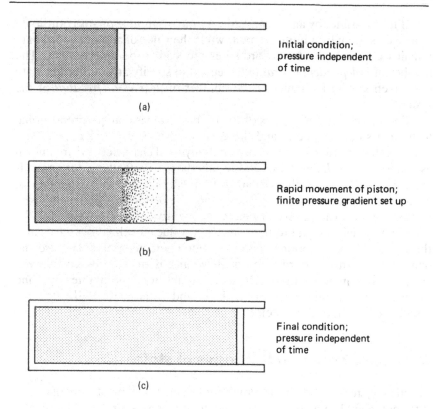

(a)

Initial condition;
pressure independent
of time

(b)

Rapid movement of piston;
finite pressure gradient set up

(c)

Final condition;
pressure independent
of time

Figure 2.2 *A non-quasistatic expansion of a gas*

gradient rapidly disappears and, if there are no further changes in the surroundings, the system will then show no macroscopically observable changes of pressure with time, as in Figure 2.2(c). In this condition, in which the bulk physical properties of the system are well-defined and, where appropriate, are uniform throughout the system, so that each coordinate has a single value that does not change with time, the system is said to be in *thermodynamic equilibrium* and its state is called an *equilibrium state*.

The basis of the thermodynamic description of the behaviour of matter is the experimental result that a given set of values for the coordinates of a particular system specifies a unique state of that system, and that state is independent of the way in which it was achieved. For example, it is a matter of experience that a fixed mass of a simple fluid in equilibrium is specified completely (apart from its shape) when its pressure and volume are known, assuming no long-range field effects, surface effects, etc. When the pressure and volume of this fluid take up specified values, the density, refractive index, relative permittivity, and so on, all have fixed values irrespective of the previous treatment of the mass of fluid.

If it is possible, by any means, to vary the magnitude of any coordinate of a given system by a finite amount, while the value of another coordinate remains constant, the two coordinates are said to be independent. The number of *independent coordinates* needed to specify an equilibrium state of a given system is known as the number of *degrees of freedom* of the system.

For a given system some coordinates have values that are proportional to the mass of the system, and these are known as *extensive coordinates*. All the other coordinates have an essentially local character and are known as *intensive coordinates*. To decide whether a particular coordinate is extensive or intensive, the system is notionally divided into a number of parts. Intensive coordinates are those whose value for each part is the same as that for the whole. A coordinate is extensive when its value for the complete system is equal to the sum of the values for the parts into which the system has been notionally divided. For a gaseous system pressure and density are intensive coordinates, while volume is an extensive coordinate. When a system is in thermodynamic equilibrium, both intensive and extensive coordinates are well-defined and intensive coordinates are uniform throughout the system.

2.2 Equilibrium and changes of state

When a system in an initial equilibrium state is made to change to a different equilibrium state, it is said to undergo a *process*. The states through which the system passes when it undergoes a process may be described by the coordinates of the system if, at all stages of the process, the system is infinitesimally close to a state of thermodynamic equilibrium. Under this condition each coordinate is well-defined and has a single numerical value at each instant during the process. Such a process is, effectively, a succession of equilibrium states and is termed a *quasistatic process*.

To be quasistatic, a process must be carried out so slowly that gradients in the intensive coordinates are always less than infinitesimal. All real processes are, therefore, strictly non-quasistatic, but just how slowly a process must proceed to be effectively quasistatic depends on the time, t_R, that a system needs to regain an equilibrium state when it is suddenly disturbed. When the process involves times that are much longer than t_R, the process will be approximately quasistatic. A gas at s.t.p. returns to an equilibrium state, after suffering a change in volume, in a time of about 10^{-3} s. Therefore, a process in which the change in volume takes a time of 10^{-1} s is quasistatic to a good approximation. At the microscopic level this means that, for gas contained in a cylinder by means of a piston, the speed at which the piston moves must be much less than the average speed of the molecules. Then, pressure gradients within the gas will always be less than infinitesimal; at all stages in the process the gas is infinitesimally close to a state of thermodynamic equilibrium and the state of the gas is described sufficiently closely by the parameters used to describe the state of thermodynamic equilibrium.

Since each equilibrium state of a system corresponds to definite values of the thermodynamic coordinates, an equilibrium state may be represented on a diagram of coordinates, known as an *indicator diagram*. For example, the quasistatic expansion of a gas from an equilibrium state i to an equilibrium state f may be represented by a line on the graph of pressure, p, against volume, V, as in Figure 2.3. Such a representation is not possible for a non-quasistatic process, since the coordinates are not then descriptive of the system.

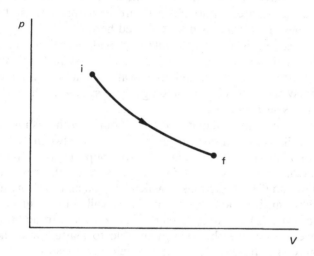

Figure 2.3 *The indicator diagram for a quasistatic expansion of a gas*

The complete sequence of equilibrium states through which a system passes during a quasistatic process is called the *path* of the process, the particular path depending on the nature of the process. A system can, of course, pass from an initial equilibrium state to a final equilibrium state by means of a process that is not a succession of equilibrium states. Then, at each instant, the coordinates will not have definite values applying to the system as a whole, the process cannot be described in terms of the coordinates and the path is, therefore, indeterminate. Classical thermodynamics deals only with equilibrium states and, therefore, can only discuss quasistatic processes, which may be regarded as limiting situations when non-equilibrium vanishes.

2.3 Boundaries and interactions

Experiment shows that the equilibrium states in which a system may exist depend on the nature of the surroundings and, in particular, on the surroundings in contact with the system, since these latter govern the interactions that can take place across the system boundary. Gaseous and liquid systems are usually placed in containers, and the nature of the wall of the container — its material and thickness — largely determines the interactions that can occur between the system and the remainder of its surroundings. From this observation has developed the concept of the *thermodynamic wall*. This is a wall, having specified properties, that surrounds the system and the inside surface of which constitutes the boundary of the system. The wall itself is, of course, part of the surroundings. Conventionally, however, the boundary is then spoken of as having certain properties, although these are really properties of the wall of the container. This usage will be followed here.

For any system it is usually possible to provide a wall of such a nature that there is practically no interaction between the system and its surroundings, and this is readily idealised to a wall that allows no interaction between the system and its surroundings. The system is then said to be *isolated* from its surroundings.

A system that is not isolated, but has a boundary that is impermeable to matter, is called a *closed system*. The interactions between such a system and its surroundings may be divided into two groups: (1) those interactions in which unbalanced Newtonian forces act between the system and its surroundings, and (2) interactions in which Newtonian forces are absent or are in equilibrium. Interactions of type (1) are called *work interactions* [3] and those of type (2) are termed *thermal interactions*. In addition, some systems have boundaries that are permeable to matter, and then mass transport may take place between the system and its surroundings. This interaction will be called a *chemical interaction,* even when a chemical reaction does not take place as a result of the mass transfer. A system with any part of its boundary permeable to matter is called an *open system.*

To illustrate the different types of interaction, consider a system consisting of a cylindrical iron rod, the system boundary coinciding with the outer surface of the rod. In a dry atmosphere the rod is a closed system whose length may be increased in the following ways: (a) by the application of an axial tension; (b) by being placed in the flame of a Bunsen burner; (c) by being placed in a magnetic field.

While the extension in situation (a) is taking place, there is a work interaction between the system and its surroundings, but note that the interaction ceases once the extension ceases. There are no forces present in situation (b). An interaction takes place so long as the rod expands but no work is done (neglecting the effect of pushing back the atmosphere). Therefore, the interaction is thermal. In situation (c) the rod lengthens by the process known as magnetostriction. This is, less obviously, a work interaction, although this should become clear through the discussion in Chapter 4.

If the iron rod is placed in an atmosphere that allows it to rust, a chemical interaction occurs. In this situation the system becomes an open one, although care is needed in specifying the position of the boundary.

2.4 Thermodynamic equilibrium

An isolated system reaches a condition, after a sufficiently long time, in which no further changes occur in the coordinates and there is thermodynamic equilibrium.

Let two such systems, separately in equilibrium, be placed in contact through a wall that allows a thermal interaction to take place. Such a wall is called a *diathermic* wall, and the two systems are said to be in *thermal contact*. In general, the result of such contact is that the values of the coordinates of the two systems change until an equilibrium state of the combined system is reached. When this happens, the two systems are said to be in *thermal equilibrium*.

The rate at which these changes occur depends on the thickness of the wall and the nature of the material of which it is made. Some materials allow only very slow rates of change of the values of the coordinates. This behaviour may be extrapolated to the postulation of an ideal material for which the rate of change is zero. Then two systems, separately in equilibrium, will remain in equilibrium in the same states, for all sets of values of their respective coordinates, when placed in contact through such a wall. A wall or boundary having this property of preventing thermal interaction is called an *adiabatic* wall. An adiabatic process is one performed by a system that is totally enclosed by an adiabatic wall or boundary. For a closed system the only interaction between a system and its surroundings in an adiabatic process is a work interaction.

Even when a closed system is in thermal equilibrium with its surroundings, it may still suffer a change of state because of unbalanced forces between the system and its surroundings and, possibly, within the system itself. Under these conditions a work interaction must occur to produce a system of balanced forces. When the forces are balanced or have become zero, the system is said to be in *mechanical equilibrium*. If a system is enclosed by a wall made of a stiff material, the effect of forces arising in the surroundings is largely to set up stresses in the wall, without much deformation and, consequently, with little effect on the system. The limiting case is the *ideally rigid* wall, which eliminates any work interaction from Newtonian forces in the surroundings, if the effects of gravity are neglected.

A closed system in thermal and mechanical equilibrium can still undergo a change of state if changes in the internal structure take place, such as chemical reactions or interdiffusion of unlike atoms to give changing local concentrations. The system is said to be in *chemical equilibrium* when the rate of such changes is zero — that is, when the macroscopic internal structure of the system does not change with time.

When the conditions for thermal, mechanical and chemical equilibria are all satisfied for a closed system, none of the coordinates changes its value with time and the system is said to be in thermodynamic equilibrium. In such equilibrium all the coordinates are well-defined and the intensive coordinates are completely uniform throughout the system.

The conditions necessary to achieve isolation of a system can now be specified more completely. For isolation, the boundary of the system must be impermeable to matter to prevent a chemical interaction, it must be rigid to prevent a work interaction (assuming that no long-range fields are present) and it must be adiabatic to prevent a thermal interaction.

Two further comments need to be made.

1 Classical thermodynamics has nothing to say about the microscopic structure of a system, but it should be noted that, when thermodynamic equilibrium is achieved, the ions, atoms or molecules of the system are still in a state of continuous agitation.

2 The presence of long-range forces causes difficulties in elementary treatments of thermodynamics. For example, gravitational forces produce non-uniform densities, so that large bodies cease to be sensibly homogeneous. Such effects will be ignored in subsequent chapters.

■ EXAMPLE

Q. A certain system consists of a fixed mass of a simple (viscous) fluid that is being forced around a closed pipework arrangement at a finite rate by a pump. When conditions are steady, is the system in a state of thermodynamic equilibrium?

A. For a system to be in thermodynamic equilibrium, it must be in a state which is completely uniform and which does not vary with time. In a simple fluid, for example, there must be no significant pressure gradient. When such a gradient is present, the system is not in thermodynamic equilibrium, even though the state may be steady.

For the system considered, the circulation at a finite rate requires a finite pressure gradient in the fluid. Therefore, although the pressure at any point does not vary with time, since the conditions are steady, the fluid is not in thermodynamic equilibrium, since the pressure is not completely uniform.

■ EXERCISES

1 Classify the following systems as open, closed or isolated.

(a) The system is a mass of gas in a container with rigid, impermeable, adiabatic walls.

(b) The system is a mass of gas in a container with rigid, impermeable, diathermic walls.

(c) The system is a sugar solution enclosed by a membrane permeable only to water and is immersed in a large container of water.

2 Classify the states in which the following systems find themselves as equilibrium or non-equilibrium states.

(a) The system is a Daniell cell on open circuit (see Section 5.5).

(b) The system is a sphere which is falling through a viscous liquid and has reached its terminal velocity.

(c) The system is a mass of gas which has remained in a container with rigid, adiabatic walls for a long time.

(d) The system is a supersaturated solution of copper sulphate into which a small crystal of copper sulphate has been placed. The container has rigid, adiabatic walls.

3 Classify the following processes as quasistatic or non-quasistatic.

(a) A mass of gas contained in a cylinder by means of a frictionless piston being compressed very slowly.

(b) A mass of gas contained in a cylinder by means of a frictionless piston being compressed very rapidly.

(c) A mass of gas contained in a cylinder by means of a rough piston being compressed very slowly.

(d) A cylinder having rigid, adiabatic walls and containing a certain mass of gas is connected by a very short tube, fitted with a stopcock, to an identical cylinder that is evacuated. The process is the passage of gas when the stopcock is opened.

4 Classify the following statements as true or false.

(a) Density is an extensive coordinate.

(b) A system enclosed by an impermeable, adiabatic boundary is necessarily in thermodynamic equilibrium.

(c) A work interaction is impossible in an open system.

(d) A thermal interaction implies that the system boundary is permeable.

(e) An isolated system is an adiabatic system, and conversely.

Notes

1   It is the role of kinetic theory to interpret these properties in terms of the forces acting on the molecules and the motions of the molecules of which matter is composed. A simple example of this kind of treatment is given in Chapter 17.

2   There is a possibility of confusion here, since, in other branches of physics, the term 'state' is frequently used to describe the different modes of aggregation of the molecules, atoms or ions of which the system is composed — that is, the conditions of solid, liquid and gas. In thermodynamics these different aggregations are called *phases*.

3   The concept of a work interaction will be discussed more fully in Chapter 4.

3

Temperature

In the previous chapter the interactions between a closed system and its surroundings were classified under two headings: (1) those in which unbalanced forces existed between the system and its surroundings, and (2) those where there were no such unbalanced forces. An interaction of the second type is a thermal interaction and will be examined in more detail in this chapter. Interactions of the first type, known as work interactions, will be considered in the following chapter.

3.1 The second and zeroth laws of thermodynamics

Experience shows that the equilibrium states achieved by a closed system, in the absence of unbalanced forces, are related to the primitive sensations of hotness and coldness and, in particular, to differences in hotness and coldness between the system and its surroundings. For example, consider a system consisting of a fixed mass of gas such as nitrogen contained in a cylinder with diathermic walls by means of a piston, as in Figure 2.1(b). The state of this system is uniquely specified by the values of the pressure p and the volume V. If a careful set of measurements of the equilibrium values of p and V is made in the open on a day in winter, the plot marked W is obtained (see Figure 3.1), whereas repeating the measurements on the same system on a day in summer would give the plot S. The difference is not the result of inaccuracy in the measurements or of a neglected work interaction, but is a consequence of the difference in hotness of the surroundings when the measurements are made.

The recognition of hotness is one of primitive experience, but physiological sensations are rather unreliable, as they depend on the immediate past experience of the sensing organ. However, the concept of hotness may be refined by relating it to some mechanical operation. Most materials suffer a change in density with a change in hotness when maintained at constant pressure, provided that a phase change does not occur. Excluding

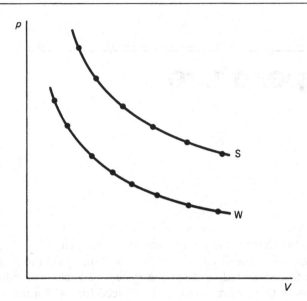

Figure 3.1 *Graphs of pressure* p *and volume* V *for a fixed mass of gas in winter (W) and summer (S)*

for the present a few materials, such as water, that behave differently from the majority, and assuming that no phase changes occur, *hotter* will be defined to mean less dense and *cooler* more dense, at constant pressure.

Consider two closed systems, A and B, that are separately in thermodynamic equilibrium, and let them be placed in thermal contact. In general, both systems will suffer a change in hotness. If thermal contact between systems A and B causes system A to become cooler (i.e. more dense at constant pressure), then system B becomes hotter, and system A is then defined to be hotter than system B. This relationship may be denoted

$$A > B$$

When the thermal contact is maintained for a sufficiently long time, thermal equilibrium is achieved. The hotness of system A is then defined to be the same as that of system B, the condition being denoted by

$$A = B$$

This result does not depend on the nature or position of the thermal contact.

The observation described in the above paragraph is universal for pairs of closed systems when there is only a thermal interaction and may be expressed formally as follows.

When thermal contact between two closed systems, A and B, causes B to get hotter and A to get colder, no matter where that contact is made, there is no process that can cause A to get hotter and B to get colder which does not involve a work interaction.

This generalisation of experience is one statement of what is known as the second law of thermodynamics [1]. The second law allows a consistent ordering, in terms of hotness, to be made for any two closed systems that are in mechanical and chemical equilibrium. Further, it follows that, if the ordering in terms of hotness of three closed systems, A, B and C, is represented by

$$A > B > C$$

then C cannot be hotter than A — that is, if $A > B$ and $B > C$, then $A > C$. Therefore, a one-dimensional sequence exists on which all closed systems may be placed in order of hotness.

In the limit, when the systems A, B and C have been in thermal contact for a sufficiently long time, so that thermal equilibrium has been achieved, all three systems have the same hotness and this condition may be represented by the statement:

$$\text{when } A = B \text{ and } B = C, \text{ then } A = C$$

or, in words, two closed systems separately in thermal equilibrium with a third system are in thermal equilibrium with each other. This statement is known as the zeroth law of thermodynamics, but is really a corollary of the second law.

3.2 The concept of temperature

When the density of each phase of a system is uniform, the hotness is uniform throughout the system. The zeroth law can then be given a numerical representation by introducing a scalar quantity that specifies the position of the system in the sequence of hotnesses. This property of a system which specifies its relative hotness is called its *temperature*. The temperature of a system is clearly a large-scale property, applying to the system as a whole, and the numerical value allotted to the temperature may be used in the specification of the equilibrium states of a system.

For a system with an adiabatic boundary the condition for thermal equilibrium is that the temperature throughout the system must be uniform. However, when the boundary is diathermic, the zeroth law indicates that, in addition, the temperature of the system must be the same as that of the surroundings. Temperature can then be given the following operational definition:

The temperature of a closed system is the property that determines whether or not the system is in thermal equilibrium with its surroundings.

When a system has two or more independent coordinates, it is possible to change the state of the system while keeping the temperature constant. Consider two closed systems, A and B, each of which has two independent coordinates, X_A, Y_A and X_B, Y_B, respectively. Let the systems be in thermal equilibrium when the coordinates have the values X_A', Y_A' and X_B', Y_B'. This means that, when the coordinates of the systems have these respective values, no changes occur when the systems are placed in contact through a diathermic wall. Further, since the systems are in equilibrium, each coordinate has a single numerical value characterising the system as a whole. Therefore, the equilibrium state of each system may be represented on the appropriate indicator diagram by a point, as in Figure 3.2(a,b).

Now let the value of X_A be changed to X_A'', while the values of X_B and Y_B remain X_B' and Y_B', respectively. Changes will occur in system A and a new equilibrium state will be attained in which Y_A has the value Y_A''. This new equilibrium state may be represented by a point with coordinates X_A'', Y_A'' in Figure 3.2(a). This procedure may be repeated and a series of equilibrium states of system A obtained, each of which is in thermal equilibrium with the equilibrium state of system B specified by the values X_B', Y_B', of its coordinates. From the zeroth law it follows that each of the states X_A', Y_A'; X_A'', Y_A''; . . . is in thermal equilibrium with every other state — that is, the temperature of each of these states has the same value. On the indicator diagram (Figure 3.2a) the locus of points representing all such equilibrium states corresponding to a single value of the temperature of the system gives a curve that is called an *isotherm*. Experience shows that, usually, at least part of any isotherm is a continuous curve, as in Figure 3.2.

In a similar manner it is possible to obtain a series of equilibrium states of system B, specified, respectively, by values of the coordinates X_B', Y_B'; X_B'', Y_B''; X_B''', Y_B'''; . . . , in each of which system B is in thermal equilibrium with the one state of system A having coordinates X_A', Y_A'. The locus of these equilibrium states on the indicator diagram (Figure 3.2b) is an isotherm for system B.

It follows from the zeroth law that all the states corresponding to points on the isotherm for system A are in thermal equilibrium with all states corresponding to points on the isotherm for system B. These two isotherms correspond to the same temperature of the two systems and are called *corresponding isotherms*. Other pairs of corresponding isotherms for the two systems A and B may be obtained by having different starting conditions for system A. In like manner, when one system is chosen as a standard, the corresponding isotherms of all other systems may be determined.

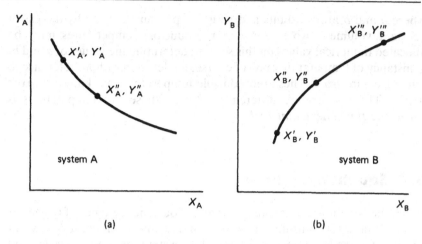

Figure 3.2 *Isotherms of two different systems*

3.3 Empirical temperature

For quantitative work it is necessary to assign numbers to the isotherms of a system, corresponding isotherms of all systems being given the same number. This process is called setting up a *temperature scale*. The number allotted to any particular set of corresponding isotherms is the value of the temperature, symbol θ. At this stage, one system will be chosen as a standard and rules agreed for numbering its isotherms under certain conditions, or constraints. The standard system, with its specified constraints, is called a *thermometer*.

For the chosen standard system, under the specified constraints, a number is allotted to each isotherm. This numbering may be done at will, but the function used, which need not have a simple analytical form, must retain the order of hotness given by the zeroth law — that is, the function chosen must be monotonic with respect to hotness. Because of this element of choice in selecting the standard system and in labelling its isotherms, θ is known as the *empirical temperature*.

In conventional thermometry constraints are applied to the standard system so that only one independent coordinate changes its magnitude with change in temperature. For example, if a fixed mass of gas is chosen as the *thermometric substance,* it is convenient to maintain this at constant volume and to use changes in the pressure exerted by the gas to detect changes in temperature [2]. It is customary to use a linear function to label the isotherms, so that empirical temperature on the appropriate *constant-volume gas scale,* denoted by $\theta_{g,V}$, is defined by the equation

$$\theta_{g,V} = a\, p_V + b \qquad (3.1)$$

where a and b are constants and p_V is the pressure exerted by the gas at constant volume. Two steady and reproducible temperatures must be allocated numerical values on this scale to determine the values of a and b. Constancy of temperature may, of course, be detected without reference to a temperature scale. These reproducible temperatures are known as *fixed points*. The temperature difference between these two temperatures is called the *fundamental interval*.

3.4 Equations of state

When the labelling process has been carried out, the existence of isotherms ensures that a relationship can be written down between the independent coordinates needed to specify the equilibrium states of a chosen system and the empirical temperature θ. For a system with independent coordinates X and Y this relationship may be written

$$f(X, Y, \theta) = 0 \qquad (3.2)$$

where f is a function that must be determined experimentally for the particular system and empirical temperature scale used. Experiment shows which coordinates should be included in the above equation, and the nature of the system governs certain aspects of the relationship, but the final form is determined by the arbitrary choice of empirical temperature scale.

It follows from Equation (3.2) that each equilibrium state of the chosen system is represented by a point in X–Y–θ space and the totality of points, representing all possible equilibrium states, forms a (usually) continuous surface. Equation (3.2) is the equation of this surface. It gives the relationship between the coordinates, and is known as the *equation of state* of the system.

As an example, consider a system consisting of a fixed mass of a simple fluid. This system has its equilibrium states specified by the sets of values of pressure p, volume V and empirical temperature θ. Each equilibrium state is represented by a point in p–V–θ space, as in Figure 3.3(a), and part of the surface representing the totality of states might appear as in Figure 3.3(b). The equation of this surface is the equation of state of the system and may be written

$$\phi(p, V, \theta) = 0 \qquad (3.3)$$

where ϕ is a function that must be found by experiment. For this system, whatever temperature scale is used, the limitation on ϕ imposed by nature is that Equation (3.3) must be such that V never increases as p increases when θ is contant. Note that, in general, ϕ will not have a simple analytical form, but the equation of state does exist and the sets of coordinate values

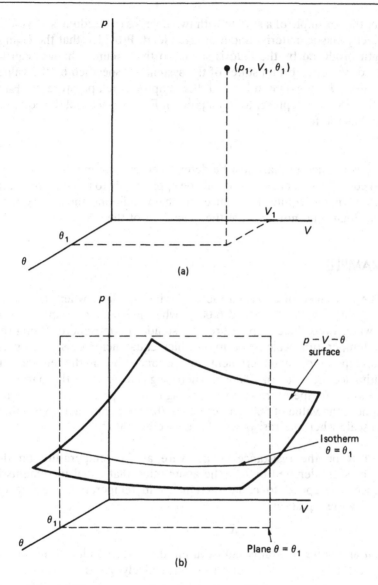

Figure 3.3 *(a) The representation of a single equilibrium state of a simple fluid in p–V–θ space; (b) the surface in p–V–θ space, showing the totality of equilibrium states for a simple fluid*

can be tabulated. Further, it must be noted that the equation of state applies only to equilibrium states of the system. The isotherm for $\theta = \theta_1$ for the system whose equation of state is Equation (3.3) is the intersection between the surface and the plane given by $\theta = \theta_1$, as shown in Figure 3.3(b).

Another example of a system with two degrees of freedom is a cylinder of perfectly elastic material under an axial load. Provided that the change in length produced by the load is small, so that volume changes can be neglected, the equilibrium states of the system are specified by the values of the load F, the length L and the empirical temperature θ. Each equilibrium state is represented by a point in F–L–θ space and the equation of state has the form

$$\psi(F, L, \theta) = 0 \qquad\qquad (3.4)$$

where ψ is a function that must be determined experimentally.

An equation of state can normally be applied only to a system in a state of thermodynamic equilibrium, since, in other conditions, the thermodynamic coordinates do not describe the behaviour of the system.

■ EXAMPLE

Q. The resistance of a certain piece of wire is 5.00 Ω when it is at the temperature of melting ice and 6.00 Ω when it is at the temperature at which water boils under a pressure of 1 standard atmosphere. Using this wire, a temperature scale is set up on which equal increments in temperature correspond to equal increments in resistance. Giving the temperature of melting ice the value of 100 deg, where deg is the unit on this particular scale, and the temperature of water boiling under a pressure of 1 standard atmosphere the value of 500 deg, calculate the temperature as determined by this scale when the resistance of the wire is 5.40 Ω.

A. Let R be the resistance of the wire at the temperature on this resistance scale denoted by θ_R. The assumption that equal increments in resistance correspond to equal increments in temperature may be expressed by the equation

$$R = a\,\theta_R + b$$

where a and b are constants. Substituting the corresponding values for R and θ_R at the two fixed temperatures, respectively, gives

$$5.00 = 100\,a + b$$
$$6.00 = 500\,a + b$$

from which

$$a = 1/400 \ \Omega \ \text{deg}^{-1}$$

and

$$b = 4.75 \ \Omega$$

Therefore, when $R = 5.40\ \Omega$,

$$\theta_R = (5.40 - 4.75)\,400$$
$$= 260 \ \text{deg}$$

■ EXERCISES

1 Discuss whether or not isotherms of a particular system for different temperatures can intersect.

2 Discuss what is meant by the temperature of an individual atom of a system.

3 Classify the following coordinates as intensive or extensive: pressure, volume, temperature, load, length.

4 Classify the following statements as true or false:

(a) For a closed system an isothermal process is one that necessarily takes place at constant density.
(b) In a closed system with coordinates pressure, volume and temperature, an isothermal process cannot take place at constant pressure.

5 An empirical temperature scale θ_R is defined so that equal changes in the resistance R of a length of pure copper wire indicate equal changes in temperature. A second empirical temperature scale θ_r is similarly defined, but using the change in resistance r of a thermistor. At the melting point of ice (the ice point) $R = 5.63\ \Omega$ and $r = 1260\ \Omega$, while at the temperature of steam under a pressure of 1 atmosphere (the steam point) $R = 8.10\ \Omega$ and $r = 20.3\ \Omega$. If, on both scales, the ice point is given the value of 100 units and the steam point the value of 1000 units, calculate the value of room temperature on both scales, where it is found that $R = 6.18\ \Omega$ and $r = 460.1\ \Omega$.

Notes

1 This form of the second law is due to F. C. Frank. Alternative statements will be considered in Chapter 7.

2 Gas thermometers are discussed more fully in Section 9.3.

4

The first law of thermodynamics

This chapter starts by looking more closely at the work interaction between a system and its surroundings. The systems usually first discussed in an elementary treatment of classical thermodynamics remain at rest in the laboratory. Therefore, when work is done on such a system, it does not suffer changes in its bulk kinetic or potential energies. A major aim of this chapter is to investigate the effects of doing work on a system when there are no changes in the bulk kinetic and potential energies.

4.1 Work and thermodynamic systems

Work, in the Newtonian sense, is done when a force moves its point of application in its own direction, the amount of work done by a force being equal to the product of the force and the component of the displacement parallel to the force. In particular, when a body of mass m is raised at constant velocity through a height h in a gravitational field in which the magnitude of the acceleration of free fall is uniform and equal to g, the work that has to be performed on the body is equal to mgh and the energy of the body increases by this amount.

This latter result suggests that a work interaction between a system and its surroundings may, in principle, be measured from the displacement, relative to some reference level, of a weight acting on the system through a suitable arrangement of ideal strings and pulleys. However, it is necessary to have a definition of a work interaction which gives a clear distinction between work and thermal interactions, as the following argument shows.

When a system becomes hotter, the effect could have been brought about by either a work interaction or a thermal interaction with the surroundings. However, in the case of a thermal interaction, although the effect in the system could have been brought about by a work input derived from a falling external weight, the effect in the surroundings, which become cooler, could not have been brought about by the reverse of such a process, involving the raising of a weight. Therefore, to allow for changes in the state of a system that are brought about simply by thermal contact, the formal definition of a *pure work interaction* in classical thermodynamics is as follows.

The interaction between a system and its surroundings (which could be another system) is a pure work interaction if, what happens in each could be repeated while the sole effect external to each is the change in height of a weight above a reference level.

On the basis of the definition of a pure work interaction, the nature of the interaction between a closed system and its surroundings can be classified as a work or thermal interaction. In making this classification the boundary of the system needs to be specified carefully. As an example, consider the situation shown in Figure 4.1(a), in which a mass of fluid F in a vessel with impermeable, adiabatic walls has a resistor R placed in it, connected to a battery B, with zero internal resistance, by means of resistanceless wires. If the fluid, the wire inside the vessel and the resistor are taken as the system, the surroundings consist of the walls of the vessel, the battery and some of the connecting wire. The system boundary then coincides with the inner surface of the vessel, and is shown dashed in Figure 4.1(a).

For the system and surroundings to interact, the switch S must be closed so that a current flows through the resistor [1]. Let the current flow for a short time and assume that the process is quasistatic. The result is that the system becomes hotter. This same process could be repeated with the surroundings replaced by a lossless dynamo D driven by a weight W falling at constant velocity, as in Figure 4.1(b). During the process the sole effect external to the system is the change in height of this weight above a reference level [2].

Similarly, the effect in the surroundings could be repeated if the system were replaced with a lossless electric motor M, as in Figure 4.1(c), that raised a weight W at constant velocity. Therefore, with the boundary of the system as in Figure 4.1(a), the interaction between the system and its surroundings is such that what happens in each could be repeated while the sole effect external to each is the change in height of a weight above a reference level — that is, the interaction is a work interaction and, in particular, is an electrical work interaction. This form of energy transfer is called *electrical work*.

However, if the system consists of the fluid F only, so that the resistor and all the connecting wire are in the surroundings, the boundary has the position shown dashed in Figure 4.1(d). Part of the boundary now coincides with the outer surface of the resistor, and this part of the boundary is diathermic. With this specification, what goes on in the surroundings when the switch S is closed cannot be repeated while the only effect external to it is the change in height of a weight above a reference level. The interaction is, therefore, a thermal interaction.

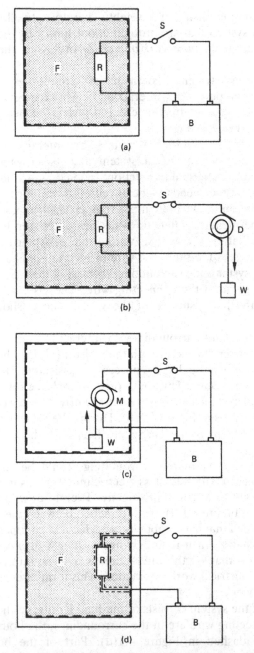

Figure 4.1 *(a)–(c) A work interaction between a system, consisting of a fixed mass of a simple fluid and a resistor, and its surroundings; (d) a non-work or thermal interaction between a system, consisting of a fixed mass of a simple fluid, and its surroundings*

4.2 Internal energy

Consider a system with an adiabatic, impermeable boundary, which has done on it work that is converted to neither macroscopic (bulk) kinetic energy of the system nor macroscopic potential energy of the system and its surroundings. Examples of such experiments are those performed by Joule in the 1840s [3]. In these experiments work was done by an external agency against frictional forces in the system, using a variety of arrangements — for example, various metals were rubbed together beneath different liquids, various liquids were forced through small holes or otherwise agitated and air was expanded or compressed. Probably the best-known experiments are those shown in Figure 4.2, in which water contained in a copper vessel is churned by a rotating paddle P while a continuous circulation of the water is prevented by the fixed vanes V, so that the water is not given bulk kinetic energy. The system is taken as the water, the paddles and vanes and the copper vessel, so that the boundary (shown dashed in Figure 4.2b) passes through the spindle S where it enters the vessel. A sensitive thermometer was used to measure the temperature rise and corrections were applied for any non-work interactions involved, so that, for an analysis and interpretation of the results, the boundary can be considered as adiabatic.

The coordinates of the system shown in Figure 4.2 are pressure p, volume V and (empirical) temperature θ. Joule's measurements showed that, for the system to go from an initial equilibrium state i, with coordinate values p_i, V_i, θ_i, to a final equilibrium state f, with coordinate values p_f, V_f, θ_f, by the performance of work only, always requires the same amount of work. This result holds even when the system does not undergo a quasistatic process, though it is then important to ensure that the work done on the system is properly determined. For example, Joule made a correction for the kinetic energy of the falling weights.

Further, not only is the same amount of work always needed to produce a given change of state under adiabatic conditions, but also the way in which the work is done does not matter [4]. It is the total amount of adiabatic work done on the system that is determined by the initial and final equilibrium states. As an illustration of what this implies, consider a system consisting of a mass of fluid in a vessel with impermeable, adiabatic walls. The work done on the system may be mechanical, as in Joule's experiments if a paddle and vanes are fitted, or it may be electrical if a resistor, connected to a suitable source of e.m.f., is included in the system. When a current I flows through the resistor under a potential difference E for a time t, the work done on the system is EIt. When the paddle wheel is caused to rotate by an applied torque τ, the work done on the system when the paddle wheel rotates through an angle α is $\tau\alpha$. For a given change of state of the system, when the only interactions are work interactions, it is

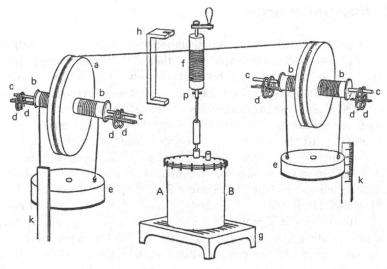

Joule's apparatus, showing driving mechanism

(a)

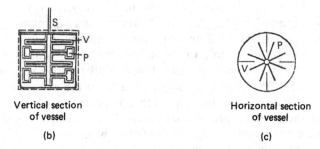

Vertical section
of vessel

(b)

Horizontal section
of vessel

(c)

Figure 4.2 *Joule's apparatus for performing adiabatic work on a fixed mass of a
simple fluid. Based on Figures 201 and 202 on pages 535 and 536 of*
A Text-book of Heat: Part II, *by H. S. Allen and R. S. Maxwell, Macmillan,
London (1945 reprint of the 1939 edition)*

found that $(IEt + \tau\alpha)$ is a constant, whatever the particular values of I, E
and τ.

The discussion so far in this section may be summarised in the
following statement.

When the state of a closed system is changed from an initial equilibrium state to a final equilibrium state solely by the performance of work,
the amount of work needed depends only on the initial and final equili-

brium states of the system, not on the means by which the work is performed nor on the intermediate stages through which the system passes.

This statement indicates that a closed thermodynamic system possesses a property, analogous to potential energy, the change in which is determined only by the initial and final equilibrium states of the system, and the measure of which is the work needed to produce the change when the boundary of the system is adiabatic. The name given to this property is *internal energy*, symbol U. When a system is brought from an equilibrium state i, in which its internal energy is U_i, to a different equilibrium state f, with internal energy U_f, by the performance on the system of an amount of work W under adiabatic conditions, the increase in the internal energy of the system is ΔU, given by

$$\Delta U = U_f - U_i = W \text{(adiabatic)} \tag{4.1}$$

Equation (4.1) gives ΔU and, therefore, defines U except for an arbitrary constant. Work that is done on the system is counted positive and produces an increase in the internal energy. What Equation (4.1) states is that when work is done on a system and is not converted into bulk potential energy or bulk kinetic energy, it becomes internal energy of the system. It follows that the performance of work on a system is to be regarded as a method of energy transfer. Once the work interaction ceases, the term 'work' is no longer needed.

Since U, for a particular system, depends only on the state of the system, it is a thermodynamic coordinate. Therefore, it is possible to specify U in terms of the independent coordinates of the system and also to use it as an independent coordinate to specify the equilibrium states of the system. For example, if the equation of state of a particular system is $f(X, Y, \theta) = 0$, in terms of its primitive coordinates, X, Y and θ, it can equally well be written $F(X, U, \theta) = 0$, and U can be expressed as a function of any two coordinates — for example, $U = U(X, \theta)$.

As U is a function of the state of a system (albeit one that is not usually known in simple analytical form), it has a exact differential, dU. Therefore, for a fixed mass of a simple fluid, which has primitive coordinates p, V, θ, the following equations are valid:

$$U = U(p, \theta) \tag{4.2}$$

$$dU = \left(\frac{\partial U}{\partial p}\right)_\theta dp + \left(\frac{\partial U}{\partial \theta}\right)_p d\theta \tag{4.3}$$

$$\oint dU = 0 \tag{4.4}$$

On the microscopic scale U is, for a system such as a simple fluid, the sum of the kinetic energies and potential energies of the atoms, molecules or ions composing the system. In other words, U is the sum of the energies associated with the translation, rotation or vibration of the individual particles.

4.3 Heat

When a system is not enclosed by an adiabatic boundary, the state of the system can be changed without the performance of work. If there is no work interaction, the system and its surroundings must be in mechanical equilibrium, so that, assuming chemical equilibrium, a change of state is produced only when the system and its surroundings are not in thermal equilibrium. In Chapter 3 it was shown that this condition means that there is a finite temperature difference between the system and its surroundings.

It is possible for a system to undergo a certain change of state by the performance of work alone and to repeat the change exactly without the performance of work. Consider, for example, the system shown in Figure 4.3, consisting of a resistor R immersed in a mass of a simple fluid F, contained in a vessel that provides a diathermic boundary. This system can undergo a change of state as a result of a work interaction with the surroundings when the weight W falls and drives the dynamo D. An

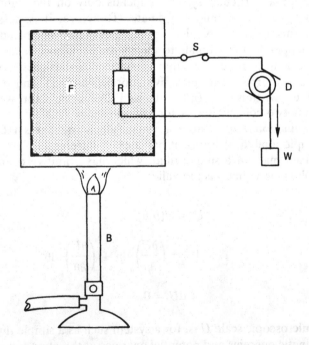

Figure 4.3 *Simultaneous work and thermal interactions*

electric current flows through the resistor R when the switch S is closed. This system can undergo the same change of state with S open, if the Bunsen burner B is lighted. This is a thermal interaction. More generally, the system of Figure 4.3 can undergo the same change of state by means of an indefinite number of processes involving different amounts of work by making use of both thermal and work interactions. For particular initial and final equilibrium states of a chosen system, the coordinates undergo changes that depend only on those equilibrium states — that is, the changes are not process-dependent. In particular, this is true of the change in internal energy of the system. However, it is not generally true that the change in the internal energy is equal to the work done on the system. Only when the process is adiabatic can the increase in internal energy be equated to the work done on the system (Equation 4.1). This does not mean that there is a defect in the concept of internal energy as defined by Equation (4.1) but, rather, that there is an energy transfer during the process which arises from the thermal interaction. This energy transfer to the system may be allowed for by replacing Equation (4.1) by

$$\Delta U = U_f - U_i = W + Q \qquad (4.5)$$

Equation (4.5) holds for all processes linking two equilibrium states of a system and defines a quantity Q, having the dimensions of energy, which is a measure of the effect of the thermal interaction. The quantity Q defined by Equation (4.5) is called *heat,* its unit is the joule and the sign convention adopted in Equation (4.5) is that heat transfer to a system, thereby increasing the internal energy, is counted positive.

The value of Q is determined by measuring the amount of adiabatic work needed to produce the same change in the state of a system: Q is never actually measured. Often, the most convenient form of work for this purpose is electrical work.

The use of the word 'heat' to describe Q in Equation (4.5) is in accord with its use in elementary physics. It can be seen that what Joule established in his experiments is that the name 'heat' is applied to the energy transfer that takes place across the boundary of a system when there is a temperature difference between a system and its surroundings.

An interesting feature of Equation (4.5) is that it involves only mechanical quantities and not the empirical temperature, θ. This gives a certain freedom to the numbering of the isotherms of a system. The sense of the one-to-one relationship between the sequence of real numbers and the sequence of systems in order of increasing hotness has not yet been agreed. It is found experimentally that, when two systems of different hotness are placed in thermal contact, the direction of heat flow, using the convention of Equation (4.5), is from hotter to colder. To conform to normal usage, the scheme adopted for numbering the isotherms will be such that hotter systems are represented by larger algebraic values of (empirical) temperature and cooler systems by smaller values. Then, if the direction of heat flow when systems A and B are placed in thermal contact is from A to B, $\theta_A > \theta_B$.

4.4 The first law of thermodynamics

Equation (4.5) is known as the first law of thermodynamics. It embodies the concept of the internal energy of a system and extends the law of conservation of energy to situations where energy transferred to a system is not converted to bulk kinetic or potential energy of the system.

The first law shows that the flow of heat and the performance of work are simply different methods of changing the internal energy of a system: both terms refer to processes [5]. Heat is the energy in transit across the boundary between a system and its surroundings because of their difference in temperature. When this temperature difference disappears, the flow of heat ceases and there is then no need for the term 'heat': the energy transferred to the system has become part of the internal energy of the system. When a system is in a particular equilibrium state, it has a definite value of U (including an arbitrary constant), but this state may have been reached in an indefinite number of ways, each involving different proportions of heat and work. Consequently, neither Q nor W is a function of the state of the system and so neither can be used as a thermodynamic coordinate.

As the above discussion shows, heat is not to be regarded as something inside a system. Nevertheless, it is convenient, and customary, to describe a thermal interaction by saying that the system absorbs a quantity of heat. The term 'absorb' is used in this book in an algebraic sense to cover the transfer of heat both to and from the system. When the direction of the heat transfer is known, and needs to be specified, the system will be said to extract heat from the surroundings when Q is positive and to reject heat to the surroundings when Q is negative.

4.5 Sign conventions

Equation (4.5) is the algebraically correct form of the first law only when a certain sign convention is followed — namely, that heat flow to, and work done on, a system are both counted positive. For the correct handling of a sign convention a number of rules must be followed and, as this is an area that sometimes causes problems, the rules are set out below.

1 A distinction must be made between the sign prefixed to an algebraic symbol in an equation or other relation and the value (positive or negative) of the physical quantity represented by the symbol [6].

2 The conventionally positive sense of all relevant physical quantities must be clearly specified.

3 All marks on diagrams, such as arrows, used with algebraic calculations must represent only the conventionally positive senses of quantities and not the sense in which these quantities are actually believed to be directed.

4.6 Reversible processes

It has been shown that neither W nor Q is a function of state. Therefore, when a closed system undergoes a process between two equilibrium states that are infinitesimally close, the first law cannot be written in the form

$$dU = dQ + dW$$

where dU, dQ and dW are all exact differentials. Rather, the first law must take the form

$$dU = q + w \tag{4.6}$$

where dU is an exact differential, because U is a well-behaved function of the state of the system, but q and w are simply infinitesimal quantities of energy transferred as heat and work, respectively. All that is known is that U is a function of the state of the system, in the sense that ΔU depends only on the initial and final equilibrium states of the system and not on the process, and may be changed by a heat transfer Q or by the performance of work W. The respective magnitudes of Q and W depend on the details of the process, but, for given initial and final equilibrium states of the system, $Q + W$ is a constant whatever the process.

Equation (4.6) will only be a useful equation if it can be integrated to give the energy transfer in a finite process. For this to be possible, q and w must be made to behave like well-behaved differentials dQ and dW, respectively. In other words, Q and W must be specified as functions of the state of the particular system under consideration and the particular process it is undergoing. Since dU is an exact differential, it is only necessary to constrain either q or w to behave as a differential, as Equation (4.6) then forces the other to behave as a differential.

One way of achieving the desired condition is to make either q or w equal to zero. When there is no work interaction,

$$w = 0 \; ; \; dU = q \equiv dQ$$

Similarly, when the process is adiabatic,

$$q = 0 \; ; \; dU = w \equiv dW$$

Since U is uniquely defined for each equilibrium state of the system (apart from an arbitrary constant), w or q is uniquely defined in the appropriate situation and may be treated as a well-behaved differential. The condition $q = 0$ is, of course, simply the condition that held in Joule's experiments that led to the concept of internal energy. Once the internal energy has been determined, the condition $w = 0$ is that used to measure q.

More important situations are those in which both q and w are non-zero. Mathematically, the sufficient conditions for a function to have a definite integral over a specified range of the argument are that the function shall be well-defined [7] and single-valued. Therefore, if w, for example, is to be treated as a well-behaved differential dW, it must be possible to express W as a well-behaved and single-valued function of the state of the system. These two aspects will now be examined.

As a particular example, consider the system consisting of an elastic rod whose length is L when subjected to a load F at a temperature θ. Each equilibrium state of the rod is uniquely specified by the equation of state, which may be written

$$\psi(F, L, \theta) = 0$$

Therefore, for any infinitesimal process from an equilibrium state, F is uniquely defined by the state of the system. Then, when the length of the rod is increased by dL while the load remains sensibly constant [8], the work done on the rod is FdL and Equation (4.6) becomes

$$dU = q + FdL \qquad (4.7)$$

q is then well-defined for the infinitesimal process and may be written dQ. However, w is not an exact differential (i.e. the differential of a function of state), because in a finite process, in which L changes from L_i to L_f,

$$W = \int_{L_i}^{L_f} FdL \qquad (4.8)$$

and, for this integral to be evaluated, F and L must be uniquely related and this relationship must be known.

Since F and L are only defined for equilibrium states of the rod, a relationship between F and L exists only for quasistatic processes, and this is the first condition that must be satisfied if W is to be a well-behaved quantity and behave as a state function.

The physical reason for this limitation can be understood when it is realised that the extension is influenced by distortions throughout the whole rod. When the magnitude of the load is changed, the extension cannot attain the equilibrium value associated with the new value of the load until all the parts of the rod have been affected by the change and the resulting distortions 'signalled' back to those places where the external forces are applied. Therefore, the equilibrium deformation cannot be attained in a time less than that needed for an elastic wave to travel from the regions where the external forces are applied to the farthest parts of the rod and back again. Even then, reflection of the elastic waves may occur — that is, the rod may be set in vibration. For the deformation process to be regarded as quasistatic, the time taken to apply each increment of load must be long compared with the slowest mode of vibration of the rod.

However, the condition that the loading is quasistatic is not sufficient to ensure that F and L are uniquely related. For the rod under load this means that the path followed in quasistatic loading must be exactly retraced in quasistatic unloading subject to the same constraints. This will be the situation when the material of the rod shows perfect elasticity without hysteresis. This is the second condition that must be satisfied. Therefore, when a rod of perfectly elastic (not necessarily Hookean) material is loaded quasistatically, F and L are well-defined and uniquely related, so that the integral in Equation (4.8) can be evaluated. Further, the work done on the rod as its length changes from L_i to L_f, under given constraints, is equal in magnitude but opposite in sign to that done on the rod when its length changes from L_f to L_i under the same conditions.

A further example may be helpful at this stage. Consider a mass of a compressible fluid contained in a cylinder by a well-fitting piston, both being of material impermeable to the fluid, as in Figure 4.4. When the fluid is in an equilibrium state, its pressure is uniform throughout. Let the position of the piston, measured from the closed end of the cylinder, be x. If the piston has an area of cross-section A and the fluid exerts a pressure p on it, the force exerted on the piston by the fluid is pA. When the fluid is compressed by the piston moving an infinitesimal distance $-dx$, the work done by the piston on the fluid is $-pA dx$, provided that p remains sensibly constant. Since $dV = A dx$, an alternative expression for this work is $-p dV$. For any infinitesimal process from an equilibrium state, p is uniquely defined by the state of the system and $w = -p dV$.

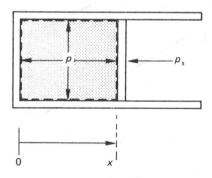

Figure 4.4 *The compression of a simple fluid*

When the fluid undergoes a finite change of volume, pressure gradients remain vanishingly small only if the piston moves sufficiently slowly. Then the pressure is uniform throughout the fluid at all stages in the process, which is, therefore, quasistatic. For a finite quasistatic change of volume from V_i to V_f, the work done on the fluid is

$$W = - \int_{V_i}^{V_f} p\,dV \qquad (4.9)$$

As with the elastic rod, W depends on the path, since p must be known as a function of V.

Consider now the effect of the frictional force between the cylinder and the piston [9]. When the pressure p exerted by the fluid is equal to p_s, the pressure exerted by the surroundings, the fluid will be in equilibrium and its state may be represented by the point I on the graph of p_s against volume V (Figure 4.5). To compress the fluid in a quasistatic manner, the value of p_s should be increased by a small amount, but this procedure simply brings the frictional force between the cylinder and piston into play. The applied pressure must rise, by a finite amount, to the value represented by A in Figure 4.5 before the piston moves. This increase in pressure corresponds to the maximum value of the frictional force, so that, when this pressure p_A is reached, a further infinitesimal increase in p_s is sufficient to cause the piston to move by an infinitesimal amount. Subsequent increments in p_s cause the gas to undergo a quasistatic

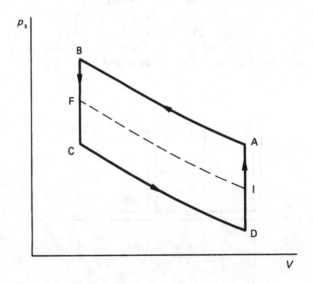

Figure 4.5 *The effect of friction on the graph of external pressure p_s needed to maintain a fixed mass of gas in a volume V*

compression, with p_s changing as shown schematically by the line AB of Figure 4.5. During this compression the pressure p of the gas in the cylinder varies as shown by the dashed line IF. To make the gas expand, p_s must be reduced, but, because of the frictional force, p_s must be reduced to the value corresponding to the point C before the piston will move. A quasistatic expansion of the gas can then be effected, with p_s varying as indicated by the line CD, while p varies as shown by FI. When the volume of the gas has the value represented by the point I, p_s can be increased to the value p_I without moving the piston. The frictional force is then reduced to zero. As a result of this complete process, in which system and surroundings are both returned to their original state, the net amount of work done by the surroundings is $\oint p_s dV$, which is equal to the area ABCDA. If the process is isothermal, this energy is transmitted to the surroundings as heat.

When the frictional force is zero, the lines DA and BC of Figure 4.5 have zero length and AB and CD coincide with IF. Then $\oint p_s dV$ is zero and at all times, provided that the process is quasistatic, $p = p_s$. In the absence of friction the work done on the gas by the surroundings during the quasistatic process IF is equal in magnitude and opposite in sign to that done by the gas on the surroundings during the process FI. Further, both the work done by the system and that done by the surroundings are determined by the same coordinates that are used to specify the equilibrium states of the system.

The foregoing discussion shows that, if W is to behave as a function of state, so that w can be treated as a well-behaved differential and be written dW, two conditions must be satisfied — namely. (1) the process must be quasistatic; (2) there must be no dissipative processes such as frictional effects or elastic hysteresis. When these conditions are satisfied, q may also be treated as a well-behaved differential, since $dU = q + w$.

Any process that satisfies the above conditions is called a *thermodynamically reversible process*, or, simply, a *reversible process*. For such a process an infinitesimal change in the applied forces will be sufficient to start the system retracing its path exactly. In a reversible process the work done on any system by the surroundings as it follows a given path from state i to state f is equal in magnitude, but opposite in sign, to that done by the system on the surroundings when it retraces the path from state f to state i. Similarly, the heat transfer on retracing the path will be equal in magnitude but opposite in sign. Therefore, the net effect of the process

$$i \xrightarrow{R} f \xrightarrow{R} i$$

where R denotes a reversible process along a specified path, is to leave both system and surroundings completely unchanged. This is an important way of testing for reversibility.

Any arbitrary closed system with two degrees of freedom may have its equilibrium states specified by coordinates X and x, where X is a generalised force and x is the corresponding generalised displacement [10]. In a reversible process w may be written dW, where

$$dW = X dx$$

or, more generally, if several different types of force X_j and corresponding displacements x_j are involved,

$$dW = \sum_j X_j \, dx_j \qquad (4.10)$$

Equation (4.6) may then be written

$$dU = dQ + \sum_j X_j \, dx_j \qquad (4.11)$$

Of the quantities in Equation (4.11), only dU is an exact differential, but by restricting processes to those that are reversible, w may be written in terms of quantities which are functions of state.

4.7 Reversible heat transfer

During a reversible process a system must always be infinitesimally close to equilibrium and, in particular, to thermal equilibrium. A flow of heat is an energy transfer under a temperature difference, and, therefore, if a heat transfer is to take place reversibly, the temperature difference causing it must be infinitesimal. If a system, initially in an equilibrium state at a temperature θ_i, is to have its temperature changed by a finite amount to a value θ_f, by absorbing a quantity of heat, this cannot be achieved in a reversible manner simply by placing the system in contact with a source of energy at a temperature θ_f. This procedure would result in a finite temperature gradient being set up in the system. Rather, the process must be carried out in a series of infinitesimal steps, so that the temperature difference between the system and its surroundings is never greater than infinitesimal.

To achieve such a condition, use is made of a series of so-called *heat reservoirs* [11]. A heat reservoir is a system of such large mass that, in any thermal interaction with another system, it suffers no appreciable change in temperature and so is able to impose its constant termperature on the system. It is assumed that this interaction does not affect the internal equilibrium of the reservoir, so that all processes within the latter are thermodynamically reversible [12].

The first reservoir in the series has a temperature θ_i and each successive reservoir has a temperature infinitesimally different from that of its neighbours, in a monotonic sequence, with the final reservoir at a temperature θ_f, as represented in Figure 4.6. Initially the system is placed in contact with the reservoir at a temperature θ_i and, after thermal

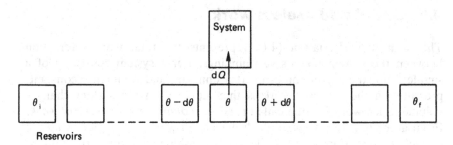

Figure 4.6 *A sequence of heat reservoirs to produce a reversible heat transfer over a finite range of temperature of the system*

equilibrium has been established, it is placed in contact with the reservoir at a temperature $\theta_i + d\theta$. While the system is in this position, it absorbs an infinitesimal amount of heat dQ reversibly and reaches an equilibrium temperature $\theta_i + d\theta$. Then the system is placed successively in thermal contact with adjacent reservoirs and allowed to come into thermal equilibrium with each before being moved to the next one. In this way the system changes its temperature reversibly from θ_i to θ_f.

4.8 **Real processes**

Finite reversible processes cannot occur in practice, as no process is entirely free from frictional or other dissipative effects nor, frequently, can the process be carried out so slowly that at all stages the system is able to adjust to the changing conditions and remain infinitesimally close to a state of thermodynamic equilibrium. The reversible process is, therefore, an ideal limit at which dissipation and non-equilibrium vanish; it is the limit of what is possible in practice. Its importance is that it is the only type of process for which exact calculations can be performed in terms of the simple description of a system and its interactions, using thermodynamic coordinates.

If either of the two conditions specified on page 37 is not satisfied, the process is termed *irreversible*. A process may be quasistatic and yet be irreversible if dissipative effects are present. Examples of irreversible processes are the following: all frictional processes; the unresisted expansion of gases; diffusive mixing of fluid streams each at different temperatures or pressures, or each of different chemical composition; the transfer of heat across a finite temperature difference. In practice, departures from reversibility are often ignored in calculations, and corrections for irreversibility are applied later.

4.9 Useful and useless work

The term w in Equation (4.6) represents the total work interaction between the system and its surroundings. For a system consisting of a simple fluid, the surroundings exert a uniform, and frequently constant, pressure on the system. Work is then done when the system changes volume. Such work is often called *pressure–volume* or *displacement work*. From an engineering viewpoint, displacement work is usually considered as *useless work,* since it is simply done in pushing back the surroundings.

However, in addition to the coupling to the uniform pressure surroundings, a system can also be coupled to a second system, which will be called the *body* and which may be treated as thermally isolated. The system can then do work on the body, termed *useful work,* as well as doing useless work in pushing back the surroundings.

Consider a system consisting of a fixed mass of gas contained within a cylinder and also coupled to a body B, as in Figure 4.7. Let the system be placed in thermal contact with a heat reservoir S and be allowed to undergo an infinitesimal reversible process at constant temperature. In this process heat q_R is absorbed by the system. If w_R is the total work done on the system, the first law may be written

$$dU = q_R + w_R$$

Since there are no frictional forces present, the condition of force balance at the piston is

$$pA = p_0A + F \qquad (4.12)$$

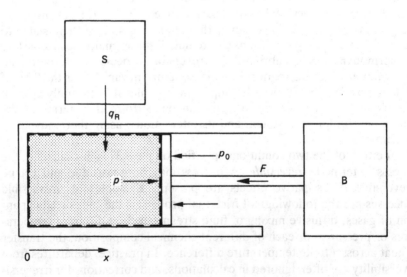

Figure 4.7 *A system of a fixed mass of gas interacting with the atmosphere and also coupled mechanically to a body B*

where p is the pressure in the gas, p_0 the pressure exerted by the surroundings, F the magnitude of the force exerted by the coupled system or body and A the area of cross-section of the cylinder.

If the piston moves a distance dx during the process, the volume of the system changes by $dV = A dx$ and Equation (4.12) can then be written

$$p dV = p_0 dV + \frac{F}{A} dV \qquad (4.13)$$

The work done on the system is $-p dV$ and, from Equation (4.13), it follows that this is made up of the pressure–volume work and the work done by the body on the system. This latter work is w'_R, where

$$w'_R = -\frac{F}{A} dV \qquad (4.14)$$

When w'_R is negative, it represents work done by the system on the body and is useful work.

■ EXAMPLES

Q1. If a liquid in a cylindrical vessel is set into bulk rotation and then left, it eventually comes to rest under the action of viscous forces. Take the liquid and the vessel as the system and assume that the walls of the vessel provide a boundary that may be treated as rigid and adiabatic.

(a) What work is done on the system as the liquid is brought to rest?
(b) How much heat is absorbed by the system as the liquid is brought to rest?
(c) Is there any change in the internal energy of the system as the liquid is brought to rest?

A1. (a) Since the boundary of the system is rigid and there are no auxiliary fittings such as resistors, there can be no work interaction between the system and its surroundings.
(b) The boundary of the system is adiabatic, so that there is no thermal interaction and, therefore, no absorption of heat by the system.
(c) The stirring process that sets the liquid into rotation gives the liquid some bulk kinetic energy ΔE_k in addition to any work that is used to increase the internal energy. The equation

$$\Delta U = Q + W$$

applies only in situations where the system does not receive any bulk kinetic or potential energy. More generally,

$$Q + W = \Delta U + \Delta E_k + \Delta E_p$$

where E_p is the bulk potential energy. In the situation described, $Q + W$ is zero while the liquid is coming to rest and so, during this process,

$$\Delta U + \Delta E_k + \Delta E_p = 0$$

or

$$U + E_k + E_p = \text{constant}$$

Because of the viscous force, E_k disappears and becomes internal energy of the system.

Q2. A mass of a certain fluid is contained in a cylinder by means of a frictionless piston and occupies a volume of 2.5 m³ when its pressure is 1.2×10^5 Pa. The fluid is then compressed reversibly under conditions in which the product of pressure and volume remains constant. Calculate the work that must be done to reduce the volume of the fluid to 1.1 m³.

A2. If p is the pressure exerted by the fluid and V its volume, the work done on the fluid by the surroundings in an infinitesimal reversible process is $-pdV$. In a finite process the work done on the fluid is W, where

$$W = -\int pdV$$

The path of the particular process here is such that

$$pV = \text{constant} = a$$

so that

$$p = a/V$$

and

$$W = -\int_{2.5}^{1.1} \frac{a}{V} dV = -[a \ln V]_{2.5}^{1.1}$$

When $p = 1.2 \times 10^5$ Pa, $V = 2.5$ m³, so that $a = 1.2 \times 2.5 \times 10^5 = 3.0 \times 10^5$ m³ Pa. Therefore,

$$W = -3 \times 10^5 \ln \left(\frac{1.1}{2.5}\right) = -3 \times 10^5 \times (-0.821) = +2.46 \times 10^5 \text{ J}$$

The positive sign to the answer indicates that the work is done on the fluid by the surroundings.

Notes

1 The system described here is not strictly a closed system, since electrons are flowing into and out of it. However, equal numbers enter and leave the system in a given time, so that the system may be treated as closed.

2 The e.m.f. produced sets up an electric field which does work in driving electric charge through the resistor. This is in accord with the definition of electrical potential difference: when Q coulombs of positive electricity leave the boundary of an electrical source, such as a dynamo, are forced through a resistor and then enter the source after falling through a potential difference E volts, the work done by the source in making the charge flow is QE joules.

3 A brief discussion of Joule's work is given in the appendix to this chapter.

4 Direct experiments to test this latter statement have probably never been performed, but the consequences of the statement are verified in practice.

5 For this reason some authors prefer to drop the terms 'heat' and 'work' in favour of 'heating' and 'working'. This usage will not be followed here.

6 For example, if the force on a charge e in an electric field of strength E is Ee, it is not permissible to state that, if e is negative, the force is $-Ee$.

7 By well-defined is meant that, as the argument varies continuously in a given range, so does the function. Further, in this range a small change in the argument gives rise to a small change in the function.

8 The use of infinitesimals in thermodynamics is legitimate, provided that the infinitesimal represents a change in a property that is small with respect to the value that the property has for the system as a whole and yet is large compared with the effect produced by a small number of molecules. For example, if the volume V occupied by a gas is notionally divided into elementary volumes, these may be treated as infinitesimals and written dV, provided that the elementary volume is very small in comparison with V and yet very large compared with the volume occupied by a few molecules at the same pressure. Then the pressure in each elementary volume is a macroscopic quantity, equal to the pressure in the total volume, since the elementary volume contains sufficient molecules for fluctuations in arrangement, momentum, and so on, not to produce detectable variations in the measured properties of the gas.

9 Since the process is quasistatic, any pressure gradients in the fluid are less than infinitesimal, so that any effects of the fluid viscosity may be neglected.

10 A generalised force is the intensive quantity which, when multiplied by the change in the appropriate extensive quantity, gives the work interaction in a particular situation. The extensive quantity is called a generalised displacement.

11 Strictly, they are internal energy reservoirs.

12 In ordinary situations the atmosphere and oceans behave as heat reservoirs.

Appendix: The work of J. P. Joule

Joule's scientific work was started at a time when the idea of conservation of energy other than in non-friction mechanical situations was not clearly perceived and the meaning of the term 'heat' was not understood. British scientists, on the whole, accepted rather uncritically the view that heat was a substance (subtile fluid) called caloric and that the temperature of a body was determined by the amount of free caloric that it contained. In France scientists considered that the caloric model and the older view, that heat was a mode of motion of the particles of a body, were equally valid and were probably different aspects of the same underlying cause.

Joule's first investigation was concerned with possible improvements to the performance of the crude electric motors available at the time. During this work, which was completed towards the end of 1840, Joule investigated the relationship between the amount of heat 'produced' in the motor in relation to the current flowing through it. In a brief series of measurements he dipped a coiled portion of the circuit into a test-tube filled with water and measured the slight change in temperature of the water for various values of the current I and resistance R. The results showed that the rate of energy dissipation in the resistance was proportional to I^2R. An important feature of this experiment, and all of Joule's subsequent experiments, is the measurement of very small temperature variations. Joule's success depended on the careful use of the best available thermometers.

In the period up to January 1843 Joule made a study of the thermal effects accompanying the production and passage of the current in a voltaic circuit. These experiments showed a clear equivalence between each type of energy dissipation and the corresponding chemical transformation, but the phenomena of the voltaic circuit gave no clue as to the nature of heat.

Joule wrote that the voltaic circuit was 'a grand agent for carrying, arranging and converting chemical heat', but this heat could either be some substance simply displaced and redistributed by the current or arise from modifications of atomic motions inseparable from the flow of the current.

Joule's master stroke was to realise that the whole matter could be put to a crucial test by extending the investigation to currents not produced by chemical change but induced by direct mechanical effect. Accordingly, in 1843 he enclosed the revolving armature of an electromagnetic engine in a cylinder filled with a known mass of water, and rotated the whole mass for a measured time between the poles of a fixed electromagnet. The energy that produced the small temperature rise observed could only have been dynamical in origin. Further, by studying the temperature changes resulting when the induced current had a current of voltaic origin added to or subtracted from it, Joule observed an approximate equivalence between the temperature rise in the fixed mass of water and the mechanical work spent in the operation.

Following this, Joule performed a series of experiments over the period 1843 to 1845 to establish the universality of this equivalence. Joule's approach was to convert work into heat by a variety of methods and measure the conversion factor. He used friction methods in which various metals were rubbed together beneath different liquids; he forced various liquids through small holes or agitated them in other ways; he compressed air into a cylinder. These methods gave values for the conversion factor that were the same to within about 15%. Joule concluded that this result was implausible on the caloric model: the caloric would need to be present in just the right amount in all these different substances to explain the result, and this conclusion carried conviction.

The paddle wheel experiments of 1845 and 1849 were precision experiments to determine an accurate value for the conversion factor, but they were not crucial for the demise of the caloric model of heat.

■ EXERCISES

1   What are the sources of irreversibility in the following processes?

   (a) The irregular stirring of a viscous liquid in thermal contact with a heat reservoir.
   (b) The shelf ageing of a Leclanché cell.

2   A closed system is in an equilibrium state i. In a certain process it has work W done on it and absorbs heat Q, reaching an equilibrium state f. Write down an expression for the difference between the internal energies of these two equilibrium states.

   Indicate any modifications to the equation that may be necessary if nuclear fission takes place within the system.

3   Classify the following processes as reversible, quasistatic only or a sequence
    of non-equilibrium states.

    (a) A mass of gas, contained in a cylinder by means of a frictionless piston,
        is compressed slowly.
    (b) A mass of gas, contained in a cylinder by means of a rough piston, is
        compressed slowly.
    (c) A mass of gas, contained in a cylinder by means of a frictionless piston,
        is compressed rapidly.
    (d) A perfectly elastic wire is extended rapidly.
    (e) A vessel with rigid walls contains a gas at a high pressure and
        communicates to the atmosphere by means of a capillary tube fitted
        with a valve. The valve is opened sufficiently for the gas to leak slowly
        into the atmosphere.

4   A system passes from an equilibrium state i to an equilibrium state f by
    process A and returns to state i by process B. During process A the system
    absorbs 100 J of heat and does work of amount 50 J. During process B
    work of amount 20 J is done on the system. Determine the heat absorbed
    by the system during process B.

5   Two systems, A and B, isolated from their surroundings are separately in
    thermodynamic equilibrium. They are brought into thermal contact through
    a diathermic wall. If there is no work interaction between the systems, show
    that, in coming to equilibrium, the heat absorbed by system A is equal and
    opposite to that absorbed by B.

6   A system consisting of a fixed mass of gas is contained in two vessels
    constructed of a rigid material that acts as an adiabatic wall. The vessels are
    connected by a short tube containing a stopcock. Initially, all the gas is in
    one of the vessels, while the other is evacuated. When the stopcock is
    opened, gas flows into the initially evacuated vessel until the pressures in
    the two vessels are the same. Calculate the thermal and work interactions of
    the system and the change in its internal energy.

7   The equation of state of a certain material showing perfect elasticity may be
    written

    $$F = C(L - L_0)/L_0$$

    where C is a function of temperature only, L_0 is the unstretched length of a
    rod of the material and L is the length under a tension F. Calculate the
    work done on the rod when its length is increased from L_i ($\neq L_0$) to L_f
    reversibly and isothermally.

5

Some simple thermodynamic systems

In this chapter several simple thermodynamic systems, including the simple fluid and elastic rod introduced earlier, will be examined briefly. The primitive coordinates of the system will be noted, an expression written down for the work interaction in a reversible process and, where appropriate, equations derived that relate changes in the primitive coordinates through quantities characteristic of a particular system.

5.1 Closed hydrostatic systems

The simple fluid is a member of the class of systems known as closed hydrostatic systems. A closed hydrostatic system is defined as one of constant mass which exerts a uniform hydrostatic pressure on its surroundings. The term, therefore, covers pure substances (including states in which different phases coexist in equilibrium), homogeneous mixtures made up of different substances in the same phase (and, therefore, having a uniform structure throughout) and heterogeneous mixtures made up of different substances in different phases (effectively, a number of subsystems in juxtaposition) [1].

Liquids and gases can be studied by being contained in a cylinder fitted with a frictionless piston, the boundary of the system being the inner surfaces of the cylinder and piston. Experiment shows that, provided that gravitational effects are neglected, the primitive coordinates needed to describe equilibrium states of such systems [2] are pressure p, volume V and (empirical) temperature θ. Solids can be studied by immersing them in a liquid in a cylinder and changing the pressure on the liquid which exerts the same pressure on the solid, as shown in Figure 5.1 [3].

Each equilibrium state of a closed hydrostatic system is represented by a point in p–V–θ space, and the equation of the surface representing the totality of such points is the equation of state, which may be written

$$F(p, V, \theta) = 0 \tag{5.1}$$

47

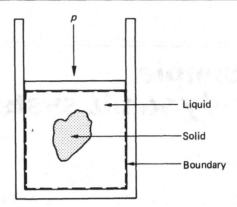

Figure 5.1   *The application of a hydrostatic pressure to a system consisting of a fixed mass of a solid immersed in a liquid*

where F is a function that must be found by experiment. Such systems, therefore, have two degrees of freedom.

When a closed hydrostatic system undergoes a reversible process, the pressure acting on the boundary has the same magnitude on the inside (produced by the system) and on the outside (caused by the surroundings). When the volume of the system changes by dV in a reversible process, the pressure remains sensibly constant. It has been shown in Section 4.6 that, in such an infinitesimal change, the work done by the surroundings on the system is given by

$$dW = -p\,dV \qquad (5.2)$$

and the first law for a reversible process may be written

$$dU = dQ - p\,dV \qquad (5.3)$$

When the process is irreversible, work may still be done on the system, but it is no longer true to write $w = -p\,dV$ for the work interaction.

In a finite reversible process, in which the volume changes from V_i to V_f, the work done on the system is given by

$$W = -\int_{V_i}^{V_f} p\,dV \qquad (5.4)$$

Equation (5.4) shows that, in a finite reversible process, W is dependent on the path, since p must be specified as a function of V: although dW is a well-defined quantity for any infinitesimal change from an equilibrium state of the system, it is not the differential of a function of the state of the system.

To find out how volume changes of a closed hydrostatic system are related to changes in pressure and temperature, Equation (5.1) may be written as an equation in V — that is,

$$V = V(p, \theta) \tag{5.5}$$

Then, in an infinitesimal reversible change in volume,

$$dV = \left(\frac{\partial V}{\partial p}\right)_\theta dp + \left(\frac{\partial V}{\partial \theta}\right)_p d\theta \tag{5.6}$$

The first term on the right-hand side of Equation (5.6) is related to the isothermal compressibility κ_θ, defined as

$$\kappa_\theta = -\frac{1}{V}\left(\frac{\partial V}{\partial p}\right)_\theta \tag{5.7}$$

and the second term is related to the cubic expansivity (strictly, the isobaric expansivity) β, defined as

$$\beta = \frac{1}{V}\left(\frac{\partial V}{\partial \theta}\right)_p \tag{5.8}$$

using an empirical temperature scale. Therefore, in general,

$$dV = -\kappa_\theta V\, dp + \beta V d\theta \tag{5.9}$$

5.2 Perfectly elastic solids

Perfectly elastic solids are systems with two degrees of freedom. The boundary of the system is the surface of the elastic solid, and experiment shows that, if the specimen is in the form of a cylinder and the load is applied axially, as in Figure 5.2, the equilibrium states of the system are specified by the values of the axial load [4] (tension or compression) F, the length L and temperature θ, provided that volume changes are negligible. All equilibrium states of the system are represented by points on a continuous surface in F–L–θ space, and the equation of this surface is the equation of state of the system.

When the length of a cylinder under a tension F increases from L to $L + dL$ in a reversible process, the work done on the system is, using the sign convention of Chapter 4,

$$w \equiv dW = F dL \tag{5.10}$$

provided that volume changes of the system can be neglected [5]. Then

$$dU = dQ + F dL \tag{5.11}$$

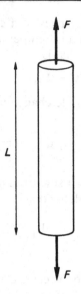

Figure 5.2   *An elastic rod of length* L *under the action of an applied axial load* F

and, in a finite reversible process, in which the length changes from L_i to L_f,

$$W = \int_{L_i}^{L_f} F \, dL \qquad (5.12)$$

The process will be reversible provided that the material is perfectly elastic and the change in length is produced slowly. Perfectly elastic means that F and L must have the same unique relation in both loading and unloading. If the load is not built up gradually in a finite process, the cylinder will undergo an accelerated motion (in fact, vibration), because unbalanced forces will be present, and it will pass through states that cannot be described by the thermodynamic coordinates.

In general, when the cylinder undergoes a reversible change in length dL,

$$dL = \left(\frac{\partial L}{\partial \theta}\right)_F d\theta + \left(\frac{\partial L}{\partial F}\right)_\theta dF \qquad (5.13)$$

The first term on the right-hand side of Equation (5.13) is related to the linear expansivity λ, defined by

$$\lambda = \frac{1}{L}\left(\frac{\partial L}{\partial \theta}\right)_F \qquad (5.14)$$

Experimentally it is found that λ depends slightly on F and more strongly on θ, but for small temperature ranges it may be treated as a constant. The second term on the right-hand side of Equation (5.13) is related to the (isothermal) Young's modulus E of the material, defined by

$$E = \frac{L}{A}\left(\frac{\partial F}{\partial L}\right)_\theta \tag{5.15}$$

where A is the area of cross-section of the cylinder. E depends on θ, but only slightly on F. For many materials at constant temperature, E is a constant, provided that the change in L is small compared with L. Then dL is proportional to dF and the behaviour is termed Hookean.

Equation (5.13) may be rewritten in terms of λ and E to give

$$dL = L\,\lambda\,d\theta + \frac{L}{AE}\,dF \tag{5.16}$$

If λ, A and E are treated as constants, Equation (5.16) may be integrated to give

$$\ln\left(\frac{L_f}{L_i}\right) = \lambda(\theta_f - \theta_i) + \frac{(F_f - F_i)}{A\,E} \tag{5.17}$$

where the initial equilibrium state has coordinates F_i, L_i, θ_i and the final one has coordinates F_f, L_f, θ_f. When F is constant, Equation (5.17) may be written

$$\ln\left(\frac{L_f}{L_i}\right) = \lambda(\theta_f - \theta_i) \tag{5.18}$$

and if $(\theta_f - \theta_i)$ is small, this becomes

$$L_f = L_i[1 + \lambda(\theta_f - \theta_i)] \tag{5.19}$$

5.3 Liquid–vapour interfaces

Consider a system consisting of a one-component liquid in equilibrium with its vapour. In the absence of any external forces, the liquid takes up a spherical shape, so achieving the smallest surface area for its mass. From this it may be inferred that, to increase the surface area of the liquid drop, it is necessary to do work on the drop. There is, therefore, an excess of energy associated with the liquid–vapour interface [6]. The measure of this excess energy is the specific surface free energy σ of the liquid, which is the work that must be done on the surface, in a reversible, isothermal process, to increase the surface area by unity. The units of σ are J m^{-2}. This excess of surface energy is equivalent to saying that the liquid surface is in a state of tension. The surface tension of a plane liquid surface is defined as the force acting parallel to the surface and at right angles to a line of unit length anywhere in the surface. For a pure liquid the surface tension (units N m^{-1}) is numerically equal to σ, which does not depend on the area of the liquid surface.

For a closed one-component system that contains a liquid–vapour interface, such as that shown in the tapered, narrow horizontal tube of Figure 5.3, the system boundary is the inner surface of the containing vessel. Each equilibrium state is specified by the values of pressure p,

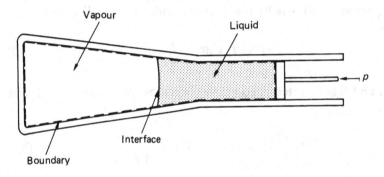

Figure 5.3 *A closed system having a liquid–vapour interface*

volume V, temperature θ, specific surface free energy σ and the area of the liquid–vapour interface A, giving four degrees of freedom. The performance of work on the system will change the volumes of the bulk phases on either side of the interface and also the area of the interface. When the work is performed in a reversible manner and the sign convention of Chapter 4 is used,

$$w \equiv dW = -p\,dV + \sigma\,dA \qquad (5.20)$$

However, when changes in volume are negligible, as in the arrangement of Figure 5.4, in which a liquid film is formed on a frame with one arm that can move without friction [7],

$$w \equiv dW = \sigma\,dA \qquad (5.21)$$

Under these conditions equilibrium states of the system are represented by points in σ–A–θ space (the system now has only two degrees of freedom) and the first law for reversible processes becomes

$$dU = dQ + \sigma\,dA \qquad (5.22)$$

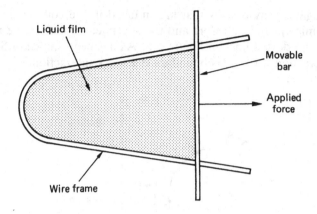

Figure 5.4   *A surface film on a framework*

5.4 Paramagnetic solids

The magnetisation M of any specimen of a magnetic material is defined as the magnetic moment per unit volume — that is,

$$M = \frac{m}{V} \tag{5.23}$$

where m is the magnetic moment of a specimen of volume V. When the magnetisation is produced by an applied magnetic field of field strength H_a (the H-field), the magnetic susceptibility, χ_m, is defined as

$$\chi_m = M/H_a \tag{5.24}$$

When the material is isotropic and homogeneous, χ_m is independent of position in the specimen and the direction of H_a, so that χ_m is a scalar. For a material that is also linear, χ_m is a constant. Paramagnetic materials show this behaviour with small, positive χ_m. The magnetisation of a specimen of a paramagnetic material of ordinary dimensions involves negligible volume change, so that equilibrium states of a paramagnetic specimen are specified by the values of the applied H-field H_a, the magnetisation M and the temperature θ.

Consider a sample of a paramagnetic material that is assumed isotropic and homogeneous; the vector quantities m, M, and H_a are then all parallel and, for simplicity, may be treated as scalars. Let the material be formed into a ring of area of cross-section A and mean radius R, having wound on it a coil with n closely and uniformly spaced turns per unit length. This arrangement eliminates demagnetising effects. A current may be established in the coil by means of a battery of variable e.m.f. Assume that the coil and connecting wires have zero resistance, so that the battery only does work against any e.m.f.s that are induced in the coil. Take as the thermodynamic system the toroid and the specimen — that is, the region enclosed by the dotted lines in Figure 5.5. As has been shown in Section 4.1, the interaction across this boundary is a work interaction.

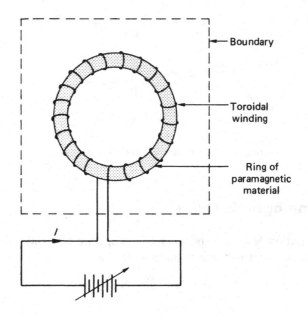

Figure 5.5 *Changing the magnetisation of a paramagnetic solid*

In the absence of a specimen, provided that R is large compared with the radius of the turns of the winding, the current I in the windings causes a uniform magnetic field H_a through them, given by

$$H_a = nI \qquad (5.25)$$

The corresponding flux density B_a, or B-field, is

$$B_a = \mu_0 H_a = \mu_0 nI \qquad (5.26)$$

where μ_0 is the permeability of a vacuum.

When the ring of paramagnetic material is introduced into the toroid, it becomes magnetised by the magnetic field produced by the coil. Since the material is assumed isotropic and homogeneous, the magnetisation is in the same direction as H_a. The B-field in the material, B_m, is given by

$$B_m = \mu_0(H_a + M) \qquad (5.27)$$

Let the current through the coil be increased by an infinitesimal amount dI in the time dt, in the presence of the paramagnetic material. The change in flux density induces a back-e.m.f. whose magnitude E is given by

$$E = \frac{d\phi}{dt} \tag{5.28}$$

where ϕ is the total magnetic flux linked with the coil.
  Now

$$\phi = 2\pi \, RnA \, B_m$$

so that

$$E = 2\pi \, RnA \, \frac{dB_m}{dt} \tag{5.29}$$

To establish the new current level in the coil, the battery must do work against the back-e.m.f. — that is, do work on the system. If the process is reversible, this work is

$$w_1 \equiv dW_1 = EI \, dt \tag{5.30}$$

since $I \, dt$ is the quantity of electricity transferred in the circuit in the time dt. Therefore,

$$dW_1 = 2\pi \, Rn \, AI \frac{dB_m}{dt} dt$$

$$= 2\pi \, RA \, H_a \, dB_m \quad \text{(using Equation 5.25)}$$

$$= VH_a \, dB_m \tag{5.31}$$

where V is the volume of the paramagnetic specimen. Using Equation (5.27) gives

$$dB_m = \mu_0 \, dH_a + \mu_0 \, dM$$

so that

$$dW_1 = \mu_0 V \, H_a \, dH_a + \mu_0 V \, H_a \, dM \tag{5.32}$$

  The first term on the right-hand side of Equation (5.32) is the work needed to change the energy stored in the field in the absence of the paramagnetic material, so that the second term is the work done in changing the magnetisation of the material. Denoting this latter work by dW:

$$dW = \mu_0 V \, H_a \, dM$$

or, since

$$M = m/V$$

$$dW = \mu_0 \, H_a \, dm \tag{5.33}$$

  To integrate Equation (5.33), m must be a single-valued function of H_a, a condition satisfied by paramagnetic materials. For the complete system of Figure 5.5 the differential form of the first law is

$$dU_1 = dQ + d(\tfrac{1}{2} V \mu_0 H_a{}^2) + \mu_0 H_a \, dm$$

Here U_1 includes the energy of the specimen and of the vacuum field. The first law for the paramagnetic material is

$$dU = dQ + \mu_0 H_a \, dm \tag{5.34}$$

5.5 Voltaic cells

When a metal rod is immersed in an electrolyte, an equilibrium condition is reached in which a potential difference between the rod and the electrolyte prevents further metal going into solution in the form of ions. The magnitude of the electrode potential, as the potential difference is known, depends on the particular metal and electrolyte. Therefore, two rods of different metals in an electrolyte give rise to a potential difference between the rods. This arrangement is known as a voltaic cell and the rods are referred to as electrodes. A voltaic cell is able to supply a current through an external circuit, the driving force coming from the chemical changes taking place within the cell. The electrode from which the conventional current flows into the external circuit is called the cathode and the other is called the anode.

For a voltaic cell to operate in a thermodynamically reversible manner, it is vital that the chemical reactions that occur within the cell when the conventional current flows from cathode to anode in the external circuit be reversed when the direction of the current flow is reversed. This condition eliminates so-called secondary cells and all voltaic cells that produce gas during their operation. However, this still leaves a number of voltaic cells that can be treated as thermodynamic systems, an example being the Daniell cell (1836). The Daniell cell has a copper electrode immersed in copper sulphate solution and a zinc electrode immersed in zinc sulphate solution, as in Figure 5.6. These solutions are separated by a porous membrane [8], and the copper electrode is the cathode, having a positive potential with respect to the zinc.

When the cell is on open circuit, no current flows but there is a slow diffusion of ions across the membrane. This ion movement can, however, be stopped by applying to the cell an external potential difference equal to the e.m.f. of the cell and of opposite sense. In this condition there are no chemical processes occurring and no work is being done, as no current is flowing. Therefore, provided that the temperature of the cell is the same as that of the surroundings, the cell is in thermodynamic equilibrium.

Equilibrium states of the cell are specified by the values of the appropriate coordinates : e.m.f. E, empirical temperature θ and the charge stored Z. The only form of work that can be done by the Daniell cell is electrical work, and this will be done reversibly when the magnitude of the current flowing is infinitesimal. This is achieved when an external

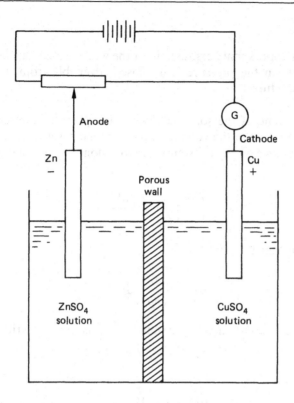

Figure 5.6 *The reversible transfer of charge in a Daniell cell*

potential difference that differs infinitesimally from the e.m.f. of the cell is applied, as in Figure 5.6. When the conventional current flow in the external circuit is from the zinc electrode to the copper one, work is done by the external source on the cell, the work dW done for a reversible transfer of charge dZ being given by

$$dW = E \, dZ \tag{5.35}$$

This process results in an increase in the 'state of charge' of the cell. Equation (5.35) shows that only changes in Z have any significance. In a finite reversible process the work done on the cell is W, where

$$W = \int_{Z_i}^{Z_f} E \, dZ = \int_0^{t_1} E \, I \, dt \tag{5.36}$$

when a current I flows for a time t_1. The first law for a reversible transfer of charge dZ from the anode to the cathode in the external circuit is

$$dU = dQ + E \, dZ \tag{5.37}$$

■ EXAMPLE

Q. Derive an approximate expression for the work done on a solid by the surroundings when the pressure is increased reversibly from p_i to p_f at constant temperature θ.

A. Let the volume of the solid be V when the pressure exerted by the surroundings is p. When the volume of the solid increases by dV, with the pressure remaining sensibly constant, the work done on the solid is $-pdV$. Now, in general,

$$dV = \left(\frac{\partial V}{\partial \theta}\right)_p d\theta + \left(\frac{\partial V}{\partial p}\right)_\theta dp$$

so that, at constant temperature,

$$dV = \left(\frac{\partial V}{\partial p}\right)_\theta dp$$

In terms of the isothermal compressibility κ_θ, this may be written

$$dV = -\kappa_\theta V \, dp$$

and, therefore,

$$dW = p \, \kappa_\theta V_\theta dp \tag{5.38}$$

To integrate this equation, the pressure dependence of both κ_θ and V must be known. Experiment shows that for crystalline solids κ_θ is very small ($\approx 10^{-11}$ Pa^{-1}) and varies little with pressure. Assume, therefore, that κ_θ is a constant and that the change in V is negligible over the range of pressure from p_i to p_f. Equation (5.38) can then be integrated to give

$$W = \kappa_\theta V \int_{p_i}^{p_f} p \, dp = \frac{\kappa_\theta V}{2} (p_f^2 - p_i^2)$$

■ EXERCISES

1 A closed hydrostatic system is taken from the equilibrium state with coordinates $p = 100$ Pa, $V = 1$ m^3 to that with coordinates $p = 4$ Pa, $V = 25$ m^3 by two different reversible processes. Using SI base units, process (a) is described by the equation

$$pV = 100$$

and process (b) by

$$p = 104 - 4V$$

Calculate the work done on the system in the two processes.

2   A vessel of volume V_0 contains n kg of gas at a high pressure. The gas is allowed to leak out slowly against the atmospheric pressure p_0, which is constant. When as much gas as possible has leaked out, show that the work W done on the gas is given by

$$W = -p_0(nv - V_0)$$

where v is the specific volume of the gas at atmospheric pressure and temperature.

3   An evacuated cylinder with rigid, adiabatic walls is connected through a valve to the atmosphere, where the pressure is constant and equal to p_0. When the valve is opened, air flows into the cylinder until the pressure there is p_0. If u_0 is the specific internal energy of air in the atmosphere, where the specific volume is v_0, and if u_c is the specific internal energy of the air in the cylinder when the air flow has ceased, show that

$$u_c = u_0 + p_0 v_0$$

4   Show that, for a closed hydrostatic system, the volume of which changes from V_i to V_f when the temperature changes from θ_i to θ_f at constant pressure,

$$V_f = V_i \exp[\beta(\theta_f - \theta_i)]$$

provided that the cubic expansivity β is a constant. If β is small, show that this expression may be written

$$V_f \approx V_i + V_i\beta(\theta_f - \theta_i)$$

5   The equation of state of a system consisting of a cylinder of a certain perfectly elastic material, subjected to an axial load F, is

$$F = C \theta(L - L_0)^2$$

where L_0 is the unstretched length, L is the length under a load F when the temperature is θ on a selected empirical scale and C is a constant. Derive expressions for the linear expansivity and Young's modulus of the material.

6   Show that, for a perfectly elastic metal in the form of a wire, having a length L at a temperature θ when under a tension F,

$$\left(\frac{\partial F}{\partial \theta}\right)_L = -EA\lambda$$

where E is the Young's modulus and λ the linear expansivity of the metal, and A is the area of cross-section of the wire.

7 Show that the pressure inside a spherical bubble in equilibrium is greater than that of the surrounding atmosphere by an amount $4\sigma/r$, where σ is the specific surface free energy of the liquid and r is the radius of the bubble.

A spherical bubble is blown, using a soap solution that has a specific surface free energy of 0.07 J m^{-2} at room temperature. Calculate the work that must be done on the surface to increase the bubble radius from 60 mm to 100 mm reversibly and isothermally.

8 Using a particular empirical temperature scale, denoted by θ, the susceptibility of a certain isotropic and homogeneous paramagnetic material χ_m is given by $\chi_m = C/\theta$, where C is a constant. A ring of the material occupies a volume V and has its magnetisation changed from M_i to M_f reversibly at a constant temperature θ_0. Show that the work of magnetisation is

$$\frac{V \mu_0 \theta_0}{2C} (M_f^2 - M_i^2)$$

where μ_0 is the permeability of free space.

Notes

1 In general, solids cannot be treated as being in states of thermodynamic equilibrium, because of the very long relaxation times of various defects and non-uniformities. However, a solid may be brought into a condition very close to that of equilibrium by careful annealing and may then be treated as a closed hydrostatic system.

2 Most thermodynamic properties of a closed hydrostatic system are largely independent of the shape of the system, so that volume is the only geometrical information that is usually given.

3 To determine the volume changes of the solid, corrections must be applied for the volume changes of the liquid and of the container.

4 Note that a state of tension is produced by balanced applied forces, as shown in Figure 5.2. A single force produces an acceleration of the system and a non-uniform load distribution within it.

5 Cylindrical specimens of practically all materials undergo a volume change when extended elastically. However, provided that the extension is small compared with the unstretched length of the specimen, the volume change is negligible.

6 For a pure liquid in contact only with its own vapour, this is usually referred to as the liquid surface.

7 The shape of the frame shown is chosen to give a stable system.

8 Originally, the gullet of an ox.

6

Some properties of gases

Gases are the simplest examples of hydrostatic systems and are also important technologically. Several important properties of gases can be determined without the aid of a calibrated thermometer, and it is the intention of this chapter to examine these properties and then formulate the qualifying properties of a model gas termed the ideal gas.

6.1 Boyle's law

The thermodynamic coordinates of a fixed mass of gas are pressure p, volume V and temperature θ. In experiments published in 1662, Robert Boyle investigated the relationship between p and V for a fixed mass of air at room temperature, using the apparatus shown schematically in Figure 6.1. The sample of air is trapped in the short, closed arm of a U-tube by a column of mercury. Assuming that this arm of the tube has a uniform cross-sectional area A, the volume of the air is given by the product of A and the length L of the air column. The pressure of the air equals that exerted by the excess column of mercury, of height h and density ρ, plus the atmospheric pressure p_0. Boyle varied h by adding mercury to the column and at the pressures involved (up to about 3 m of mercury), found that the equilibrium states of a fixed mass of air at room temperature are well represented by the equation

$$pV = \text{constant} \tag{6.1}$$

In this study it is not necessary to know the temperature of the air, but only to be able to tell whether or not the temperature changes; this is straightforward, since along an isotherm the magnitude of the thermometric property remains constant. Equation (6.1) is known as Boyle's law [1] and three representations of it are shown in Figure 6.2. Curve (a) is a rectangular hyperbola, but curves (b) and (c) are straight lines and such plots are useful for investigating the validity of Equation (6.1).

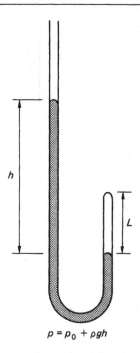

$$p = p_0 + \rho g h$$

Figure 6.1 *A schematic representation of Boyle's apparatus for the study of the behaviour of gases*

Subsequent studies extended the measurements to other gases and greatly extended the range of pressure. These experiments can be divided into two groups. Direct methods, such as those of Amagat (1870), Andrews (1869) and Holborn and Schultze (1915) were extensions of Boyle's work in that the volume of a constant mass of gas was varied [2]. A major problem in this type of experiment is that, at high pressures, the volume occupied by the gas is very small and, therefore, difficult to measure accurately. To combat this problem, indirect methods were developed by Onnes (1901) and Holborn and Otto (1926) in which the mass of gas needed to fill a definite volume at various pressures was determined.

The results of these measurements showed that, over a wide range of pressure, the behaviour of a constant mass of a gas at constant temperature is well represented by an equation of the form.

$$pV = A' + B'p + C'p^2 + D'p^3 + \ldots \qquad (6.2)$$

where A', B', C', D', ... depend on the nature, mass and temperature of the gas: for a fixed mass of a particular gas, they are functions of temperature only. The coefficients A', B', C', D', ... are of rapidly decreasing magnitude, with $A' \gg B' \gg C'$ ... at all temperatures.

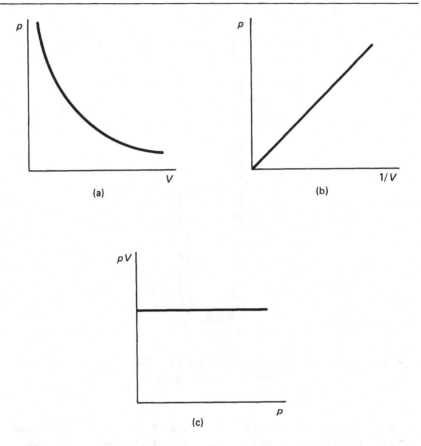

Figure 6.2 *Three useful graphical representations of Boyle's law*

Equation (6.2) shows that Boyle's law is more and more closely obeyed as the pressure of the gas is reduced. The coefficients are functions of temperature, so that, if pV is plotted against p for a fixed (arbitrary) mass of gas at different temperatures (measured on some empirical scale), isotherms of the form shown in Figure 6.3 are obtained. The convention described in Section 4.2 enables the temperatures to be classified as higher or lower: θ increases as indicated. At high temperatures pV always increases with p, while at lower temperatures pV first falls as the pressure increases and then starts to rise. In the region of the minimum on each of these latter curves there is a small range of pressure over which Boyle's law is closely obeyed. The minimum on a particular isotherm (i.e. B in Figure 6.3) is called the *Boyle point* for that isotherm. The locus of Boyle points is shown by the dotted curve through B and is described by the equation

$$\frac{\partial(pV)}{\partial p} = 0 \tag{6.3}$$

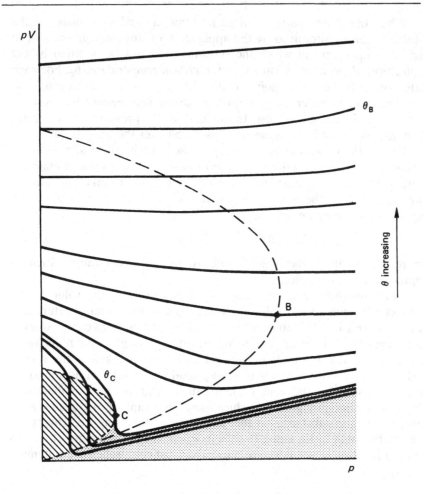

Figure 6.3 *Isotherms for a real gas*

One isotherm has its minimum at the value of p equal to zero, and shows an extended range of pressure over which pV is practically constant — that is, over which Boyle's law is closely satisfied. The temperature of this isotherm is known as the *Boyle temperature*, θ_B, and is given by the condition

$$\theta = \theta_B \quad \text{when} \quad \left(\frac{\partial(pV)}{\partial p}\right)_\theta = 0 \text{ at } p = 0 \qquad (6.4)$$

Since the coefficient B' is a very much larger than any of the coefficients except A', an alternative condition for the Boyle temperature is

$$\theta = \theta_B \text{ when } B' = 0 \qquad (6.5)$$

When the temperature is sufficiently low, a condensed phase of the substance can be produced by the application of sufficient pressure. The highest temperature at which the substance can just be liquified by the application of pressure is known as the *critical temperature,* θ_c. For each state corresponding to a point in the shaded region of Figure 6.3 the substance is in a condensed phase, while the hatched region denotes states in which liquid and gas coexist. In the limit, as the pressure tends to zero, all substances exist in the gaseous phase, whatever the temperature.

Rather than considering arbitrary masses of gas, there should be an advantage in comparing the behaviour of gaseous systems that contain the same number of molecules. One way of doing this is to take one mole of molecules [3] as the system under consideration. For such a system Equation (6.2) may be written

$$pV_m = A + Bp + Cp^2 + Dp^3 + ... \qquad (6.6)$$

where V_m is the molar volume and $A,B,C,D,...$ are the coefficients appropriate to one mole.

Equation (6.6) indicates that, as p tends to zero, the value of the product pV tends to the value of A appropriate to the temperature and nature of the gas considered — that is, Boyle's law is more closely obeyed as the pressure is reduced [4]. However, experiment gives the remarkable result that, when one mole of molecules of a gas is taken as the system under consideration, not only does pV_m tend to the value A as p tends to zero, but also A has the same value for all gases at the same temperature. This is illustrated in Figure 6.4, which shows the graph of pV_m against p for several common substances at the temperature of melting ice. Therefore, the coefficient A in Equation (6.6) is independent of the nature of the gas and is a function of temperature only. The most comprehensive statement of Boyle's law is, therefore,

$$\lim_{p \to 0} pV_m = A = f(\theta) \qquad (6.7)$$

where $f(\theta)$ is a universal function, independent of the nature of the gas.

6.2 Joule's law

When a fluid is allowed to undergo an expansion in which there is neither a work interaction nor a thermal interaction, it is said to undergo a *free expansion*. The free expansion of several gases was studied by Gay-Lussac in 1806 and the measurements on air repeated with greater precision by Joule in 1845. In Joule's early experiments a copper vessel R containing the air was joined to a similar, but evacuated, vessel E by a short pipe fitted

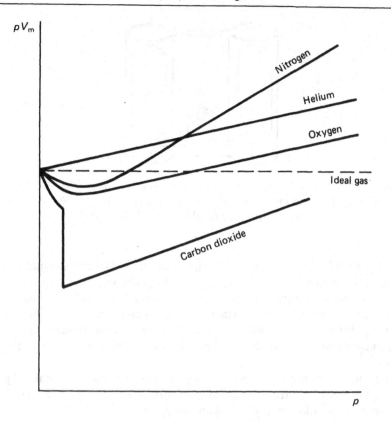

Figure 6.4 *Isotherms for 1 mol of molecules of different gases at the temperature of melting ice*

with a stopcock. The whole apparatus was immersed in a water-bath containing the minimum amount of water, the intention being to infer changes in the temperature of the air from changes in the temperature of the water (see Figure 6.5). When equilibrium was established, the water was thoroughly stirred and its temperature measured, using a sensitive thermometer. Then the stopcock was opened, so that the air expanded into the evacuated vessel until equilibrium was established. This expansion was so rapid that there was a negligible thermal interaction between the water-bath and the surroundings. Finally, the water was again carefully stirred and its temperature measured. In his rather insensitive experiments Joule found that 'no change in temperature occurs when air is allowed to expand in such a manner as not to develop mechanical power'.

In analysing this result further, care must be taken because the free expansion of a gas is an irreversible process: finite pressure gradients exist

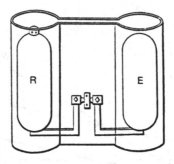

Figure 6.5 *Joule's apparatus for studying the free expansion of gases. Taken from page 24 of* The Free Expansion of Gases, *translated and edited by J. S. Ames, Harper and Brothers, New York, 1898*

in the gas while the expansion is taking place; there is a transformation of some of the internal energy of the gas into kinetic energy of 'mass motion' and this kinetic energy is then dissipated through viscosity into internal energy again. However, at the start and at the end of the process the gas is in an equilibrium state in which p, V and θ are well-defined quantities, and relationships may be looked for between the coordinate values in these states.

The constraints imposed upon the system (taken as the fixed mass of gas that undergoes the free expansion) ensure that $Q = W = 0$, so that applying the first law to a free expansion gives

$$\Delta U = 0 \qquad (6.8)$$

This equation gives no information about possible temperature changes in the system but this point can be investigated by using the fact that U is a state function. Let the system undergo an infinitesimal free expansion. The change, dU, that occurs in U may be written as a function of p and θ:

$$dU = \left(\frac{\partial U}{\partial p}\right)_\theta dp + \left(\frac{\partial U}{\partial \theta}\right)_p d\theta \qquad (6.9)$$

From Equation (6.8) the value of dU is zero; Joule's experiments suggest that, at least to a first approximation, $d\theta$ is zero; and, for there to be an expansion at all, dp is non-zero. Putting these conditions into Equation (6.9) gives

$$\left(\frac{\partial U}{\partial p}\right)_\theta = 0 \qquad (6.10)$$

(i.e. U is independent of p for a gas at constant temperature). Equation (6.10) is known as Joule's law.

Now, in general,

$$\left(\frac{\partial U}{\partial p}\right)_\theta = \left(\frac{\partial U}{\partial V}\right)_\theta \left(\frac{\partial V}{\partial p}\right)_\theta$$

and, for gases,

$$\left(\frac{\partial U}{\partial p}\right)_\theta = 0 \text{ but } \left(\frac{\partial V}{\partial p}\right)_\theta \neq 0$$

Therefore, it is also true for gases that

$$\left(\frac{\partial U}{\partial V}\right)_\theta = 0 \tag{6.11}$$

Since U can only be a function of p, V and θ, Equations (6.10) and (6.11) indicate that Joule's law may also be written in the form

$$U = \phi(\theta) \tag{6.12}$$

where ϕ is a function of temperature only.

More precise measurements of $(\partial U/\partial p)_\theta$ for gases have been made, usually by less direct methods than that of Joule. For example, Rossini and Fransden (1932) allowed the gas to leak slowly from a vessel to the atmosphere and measured the energy input needed to maintain the gas temperature constant. They found that $(\partial U_m/\partial p)_\theta$ is non-zero but is independent of pressure, depending only on temperature — that is,

$$\left(\frac{\partial U_m}{\partial p}\right)_\theta = I(\theta)$$

where $I(\theta)$ is a function of temperature only, and this may be integrated to give

$$U_m = p \, I(\theta) + F(\theta) \tag{6.13}$$

where $F(\theta)$ is another function of temperature only which is found to be the same for all gases. From Equation (6.13) it follows that, in the limit of vanishingly small pressures, U_m is a universal function of temperature only. Essentially, this is because, when the pressure of the gas is very small, that part of the internal energy which depends on the pressure becomes a very small fraction of the total internal energy. Joule's law may then be written in the precise form

$$\lim_{p \to 0} U_m = F(\theta) \tag{6.14}$$

where $F(\theta)$ is a universal function of temperature.

6.3 The ideal gas

The behaviour of gases in the limit of vanishingly small pressures is described by the simple Equations (6.7) and (6.14). A theoretical gas may be imagined which obeys these simple laws at all pressures. Such a gas is called an *ideal gas* and, by definition, satisfies the following two independent equations at all pressures:

$$\left. \begin{array}{c} pV_m = f(\theta) \\ U_m = F(\theta) \end{array} \right\} \tag{6.15}$$

where $f(\theta)$ and $F(\theta)$ are both universal functions of temperature. Both these equations were obtained without the need for a particular temperature scale: it was sufficient to know that the temperature was constant.

■ EXAMPLE

Q. A certain gas has the equation of state

$$pV_m = A + Bp$$

where A and B are functions of temperature only. Derive an expression for the work done on a system consisting of 1 mol of molecules of this gas when it expands reversibly and isothermally from an equilibrium state with coordinates p_i, $V_{m,i}$ to one with coordinates p_f, $V_{m,f}$.

A. When any closed hydrostatic system undergoes a reversible increase in volume dV at a pressure p, the work done on the system, dW, is given by

$$dW = -pdV$$

For the gas under consideration

$$pV_m = A + Bp$$

so that, in a reversible isothermal process, when A and B may be treated as constants,

$$pdV + V_m dp = Bdp$$

and, therefore,

$$-pdV = (V_m - B)dp$$

The work done on the system in the finite process is, then,

$$W = -\int pdV = \int_{p_i}^{p_f} (V_m - B)dp$$

From the equation of state,

$$V_m - B = A/p$$

so that

$$W = \int_{p_i}^{p_f} \frac{A}{p}\, dp = A \ln\left(\frac{p_f}{p_i}\right)$$

Since p_f is less than p_i, W is negative, indicating that work is done by the gas on the surroundings.

[Alternatively, direct substitution of $p = A/(V_m - B)$ into $dW = -pdV$ gives

$$W = A \ln\left(\frac{V_{m,i} - B}{V_{m,f} - B}\right)$$

which equals, of course, $A \ln (p_f/p_i)$.]

■ EXERCISES

1 A faulty barometer contains a small quantity of air in the space above the mercury. When the length of the mercury column is 75.8 cm, the space above the mercury has a length of 8.0 cm. On pushing the tube 3.3 cm into the mercury, the mercury column attains an equilibrium length of 75.5 cm. What is the pressure of the atmosphere in centimetres of mercury in these circumstances? What assumptions are made in the calculation?

2 A piston pump having a barrel of effective volume 100 cm^3 is used to exhaust a vessel of volume 2.5×10^3 cm^3. If the initial pressure of the air in the vessel is 1 atm, estimate the reduction in pressure produced by two strokes of the pump, assuming that all processes are reversible and isothermal and that air behaves as an ideal gas. Further, estimate the number of strokes needed to reduce the pressure in the vessel from 1 atm to 0.02 atm.

3 One mole of molecules of ideal gas obeys the equation $pV_m = f(\theta)$, where $f(\theta)$ is a universal function of temperature. Show that the isothermal compressibility of such a gas is p^{-1}, the reciprocal of pressure.
   Derive an expression for the isothermal compressibility of a system consisting of 1 mol of molecules having the equation of state

$$pV_m = A + Bp$$

where A and B are functions of temperature only.

4 A certain mass of an ideal gas occupies a volume of 100 cm^3 under a pressure of 10^5 Pa (approximately 1 atm) at the temperature of melting ice.

Calculate the work done on the gas when it is compressed reversibly and isothermally to a pressure of 10^6 Pa.

5   An important application of Boyle's law is the McLeod gauge, used to measure gas pressures in the range 10–10^{-6} mmHg. The principle of the gauge is to isolate a known volume of the gas at the unknown pressure and then compress it into a small known volume, when its pressure is directly measurable.

Figure 6.6 is a schematic representation of a McLeod gauge. It has a precision-bore capillary tube A sealed to a large bulb B. Close to A is a tube C of the same bore. When the mercury reservoir is raised, a sample of the gas at the unknown pressure, p_u, is trapped in A and B. The mercury

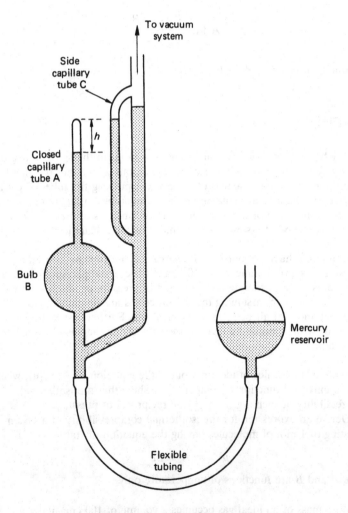

Figure 6.6   *The McLeod gauge*

reservoir is then raised until the mercury level in C reaches the same horizontal level as the top of the closed tube B. If V is the combined volume of the bulb B and the capillary A, show that p_u, in mmHg, is given by

$$p_u = \frac{h^2 A}{V}$$

provided that h is measured in mm. Here A is the area of cross-section of the tube A. Discuss the assumptions and approximations made in deriving the expression for p_u.

Notes

1 These experiments were repeated by Mariotte in France in 1676, so that Equation (6.1) is sometimes known as Mariotte's law, especially on the European mainland.

2 Amagat, for example, in some of his early experiments erected a steel tube 327 m high in a mine and filled the tube with mercury. This produced a pressure at the bottom of the tube of about 420 atm.

3 The mole is the SI unit of amount of substance, being 'the amount of substance which contains as many elementary entities as there are carbon atoms in 0.012 kg of ^{12}C'. These elementary entities must be specified and may be atoms, ions, molecules, electrons, other particles, or specified groups of such particles. The number of carbon atoms in 0.012 kg of ^{12}C is the Avogadro constant, symbol N_A, and is equal to $6.002\ 204\ 5 \times 10^{23}\ mol^{-1}$. The term 'molar' is used to mean 'divided by the amount of substance'. Although the mole is defined in terms of a number of entities, it is usually realised by weighing: one mole of atoms of element X is obtained by weighing an amount, in grams, of X equal to its relative atomic mass.

4 The product pV_m remains finite as p tends to zero, since, simultaneously, V_m tends to infinity.

7

The second law of thermodynamics

The first law of thermodynamics was developed in Chapter 4 from a study of the effects of adiabatic work on the state of a system. This law introduces the concept of internal energy (a non-primitive state function) and imposes certain limitations on the changes that can occur in a system under given constraints. For example, the first law shows that when two isolated systems, A and B, are brought into thermal contact, the heat absorbed by A is equal in magnitude but opposite in sign to that absorbed by B. However, the first law is unable to allocate the signs to these two quantities of heat. This is done by the second law, one statement of which has been discussed in Section 3.1. The second law indicates that when two isolated systems are placed in thermal contact, the temperature of the hotter one falls and that of the colder one rises, provided that phase changes do not occur in either system.

Since a difference in temperature results in a transfer of energy in the form of heat, the statement of the second law given in Section 3.1 can be rewritten: When two systems are placed in thermal contact, the direction of the energy transfer as heat is always from the system at the higher temperature to that at the lower temperature. This is the Clausius statement of the second law.

Alternative statements of the restrictions on processes that satisfy the first law are often based on observations made on real cyclic heat engines and are couched in engineering terms applicable to heat engines. A detailed study of the operation of heat engines is not necessary to understand and use the second law of thermodynamics, but a brief outline of the principles of operation will be helpful.

7.1 Heat engines

A *heat engine* is a device that does work on its surroundings as a result of a transfer of heat from the surroundings. In its simplest form it consists of a

74

system, known as the *working substance,* that can undergo thermal and work interactions with the surroundings. The range of possible working substances is very large! Many heat engines use a gas as the working substance, contained in a cylinder by means of a piston. Supplying energy to the gas causes the piston to move and work to be done on the surroundings. The thermodynamical description of the behaviour of such a device is in terms of the appropriate thermodynamic coordinates — in this case, pressure, volume and temperature. Therefore, the only processes that can be described are quasistatic processes; the action of an engine working at a finite rate cannot be described exactly in terms of the coordinates. However, as has been pointed out in Section 2.2, gases are able to respond rapidly to changes in volume, so that a heat engine using a gaseous working substance may undergo processes that are close to being quasistatic even when operating at a finite rate.

Practical heat engines are *cyclic* — that is, they repeat a regular cycle of operations continuously and, in this way, are able to do an unlimited amount of work if an unlimited supply of heat is available. An important property of any heat engine is its efficiency in converting the heat supplied into work. In evaluating this conversion it is convenient if all the work done by the engine is derived solely from the heat supplied to it. Allowance does not then have to be made for other energy changes. This condition is satisfied by engines operating in a cycle, since, at the end of each cycle, all parts of the engine, including the working substance, are in the state that they were in at the start of the cycle [1].

A study of heat engines using a gas as the working substance shows that the continuous performance of work is obtained through the alternate expansion and contraction of the working substance. Further, for the engine to do more work on the surroundings than is done on it by the surroundings in the course of one cycle of operations, the expansion must be carried out at a higher temperature than the contraction. Consequently, net work cannot be obtained from a cyclic engine without the existence of at least two bodies at different temperatures. Conversely, whenever a temperature difference exists, it is possible to use it to produce work. Many practical engines operate between what are effectively two heat reservoirs at different temperatures [2].

Observation shows that, in general, any heat engine operating in a cycle receives heat from the surroundings during some parts of the cycle and rejects heat to the surroundings during other parts of the cycle. If the working substance is treated as the system under consideration, the *thermal efficiency* η of any heat engine operating in a cycle between two reservoirs is defined as

$$\eta = -\frac{W}{Q_1} \tag{7.1}$$

Here W is the work done on the working substance in each cycle and Q_1 is the heat absorbed per cycle by the working substance from the high-temperature reservoir. W and Q_1, are of opposite sign and so a negative sign is introduced into Equation (7.1) because it is convenient to have η positive.

When a heat engine is made to run backwards, it becomes either a *refrigerator* or a *heat pump*, depending on the aims of the operation. In these devices heat is transferred from bodies of lower temperature to bodies of higher temperature by the performance of work on the working substance. When the intention is to remove as much heat as possible per cycle from the low-temperature body for a minimum work expenditure, the device is a refrigerator and the usual figure of merit is the *coefficient of performance* (C.o.P.), defined as

$$\text{C.o.P.(r)} = \frac{Q_2}{W} \tag{7.2}$$

W is the work done per cycle on the working substance and Q_2 is the heat absorbed from the low-temperature body by the working substance. W and Q_2 have the same sign.

On the other hand, when the purpose of the device is to 'upgrade' energy and transfer as much heat as possible to the hotter body for a minimum expenditure of work, it is known as a heat pump. The appropriate figure of merit is then

$$\text{C.o.P.(h)} = -\frac{Q_1}{W} \tag{7.3}$$

W and Q_1 are of opposite sign and so a negative sign is introduced into Equation (7.3).

7.2 The second law of thermodynamics

The observation that no real cyclic heat engine has an efficiency of unity is contained in Kelvin's formulation of the second law of thermodynamics, which, in modern terms, may be stated as follows.

It is impossible to devise a machine which, working in a cycle, produces no effect other than the extraction of a quantity of heat from the surroundings and the performance of an equal amount of work on the surroundings.

Despite the apparent differences between the Clausius and Kelvin formulations of the second law, the Kelvin statement may be treated as a logical consequence of the Clausius statement if it can be shown that the existence of a thermal contact that violates the Clausius statement allows a cyclic engine to be constructed that violates the Kelvin statement.

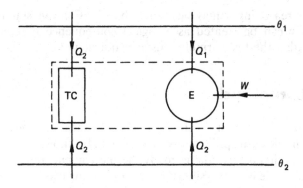

Figure 7.1 *Arrangement to demonstrate that violation of the Clausius statement of the second law implies the violation of the Kelvin statement*

Consider an engine E operating between two heat reservoirs at temperatures θ_1 and θ_2, respectively, with $\theta_1 > \theta_2$. The engine may be represented symbolically as in Figure 7.1, in which the horizontal lines represent the heat reservoirs and the circle represents the boundary around the working substance, the latter being taken as the system. TC is a thermal contact that violates the Clausius statement of the second law — that is, it allows heat to flow from a colder to a hotter body. While the engine E performs one cycle of operations in the forward sense, a quantity of heat Q_2 is transferred from the reservoir at temperature θ_2 to that at temperature θ_1 by way of TC. Let E have its amount of working substance adjusted so that in each cycle it absorbs heat Q_1 from the reservoir at temperature θ_1 and heat Q_2 from that at temperature θ_2, while the work done on it is W. Now let E and TC be combined to form a composite engine that can be treated as a system with the boundary shown dashed in Figure 7.1. In this composite engine the heat absorbed per cycle from the reservoir at temperature θ_2 is zero (remember that, for E, Q_2 is rejected), while that absorbed from the reservoir at temperature θ_1 is $Q_1 + Q_2$. Application of the first law of thermodynamics to one complete cycle of the composite engine gives

$$Q_1 + Q_2 + W = 0 \tag{7.4}$$

since the change in the internal energy of the working substance of E is zero for a complete cycle and TC is in a steady state. Since W is negative when E is operating in the forward sense, $Q_1 + Q_2$ must be positive. Therefore, the composite engine is a cyclic device that receives in each cycle a quantity of heat $Q_1 + Q_2$ from the surroundings and performs an equal amount of work on the surroundings. This violates the Kelvin statement of the second law, which may, therefore, be treated as a logical consequence of the Clausius statement.

By similar reasoning it may be shown that the Clausius statement of the second law can be treated as a logical consequence of the Kelvin statement, so that the two formulations are equivalent.

7.3 The Carnot cycle

In his discussion of the production of work by heat engines, Carnot considered a simple cyclic process that has come to be known as the *Carnot cycle*. It plays an important role in many thermodynamic arguments.

The Carnot cycle is a reversible cyclic process using two heat reservoirs at different temperatures. A convenient graphical representation of the cycle uses the appropriate generalised force and generalised displacement as coordinates. For example, when the working substance is a gas, the cycle of operations may be represented on a pressure – volume graph, as in Figure 7.2, a representation known as the indicator diagram.

In a Carnot cycle the working substance undergoes the following consecutive processes.

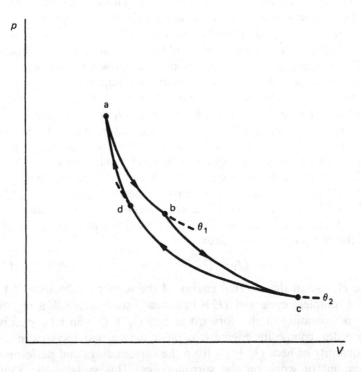

Figure 7.2 *Indicator diagram showing a Carnot cycle for a real gas. ab, cd: reversible isothermal processes. bc, da, reversible adiabatic processes*

1 The working substance at a temperature θ_1 is placed in thermal contact with a heat reservoir whose temperature is only infinitesimally higher than θ_1 and is allowed to undergo a reversible isothermal expansion. During this process the working substance absorbs a quantity of heat Q_1 from the reservoir (process ab in Figure 7.2).

2 The working substance is then placed inside an adiabatic boundary and allowed to perform a reversible adiabatic expansion until its temperature falls to θ_2, that of the other heat reservoir (process bc in Figure 7.2).

3 Next the working substance is placed in thermal contact with the reservoir at a temperature θ_2 and undergoes a reversible isothermal compression so that its temperature is never greater than θ_2 by more than an infinitesimal amount. During this process the working substance absorbs a quantity of heat Q_2 from the reservoir (process cd in Figure 7.2).

4 The reversible isothermal compression is stopped at such a state that, when the working substance undergoes a reversible adiabatic compression, it reaches its initial state when its temperature has risen to θ_1 (process da in Figure 7.2).

For an engine to be able to operate in a Carnot cycle, it must consist of the working substance, two heat reservoirs at different temperatures and a purely mechanical device capable of storing and releasing energy without introducing any thermal effects. In one complete cycle the change in internal energy, ΔU, of the working substance is zero, so that, if Q is the total amount of heat absorbed per cycle from both reservoirs and W is the work done on the working substance per cycle, the first law becomes

$$Q + W = 0 \qquad (7.5)$$

where

$$Q = Q_1 + Q_2$$

Since the Carnot cycle is thermodynamically reversible, when it is performed in the reverse sense, the signs of the energy transfers are reversed but the magnitudes are the same as when the cycle is performed in the forward sense. Any cyclic engine that operates in a reversible manner between two reservoirs is a Carnot engine, since those parts of the cycle where there is no transfer of heat must, of necessity, be reversible adiabatic processes.

This discussion of the Carnot cycle assumes that the path of a reversible adiabatic process and of a reversible isothermal process starting from a given state of a system are both uniquely defined, an assumption that needs justification. Reversible isothermal processes have θ constant, and this condition is sufficient to ensure the uniqueness of a path through a given state, since it defines a line on the representation surface of equilibrium states (see, for example, Figure 3.3b). Adiabatic processes, on the other hand, are defined by the condition $Q = 0$. Therefore, the assumption that such processes are uniquely defined amounts to assuming that there is some function of state which is constant for reversible processes for which Q is zero. Now Q is not a state function, so that the condition $Q = 0$ does not of itself define a unique path on the surface representing the equilibrium states. However, it is found experimentally that reversible adiabatics are uniquely defined and this justification will be adopted until a further development (Section 8.2) shows it to be unnecessary [3].

Strict adherence to the use of the symbols W and Q as defined in Chapter 4 has much to commend it and is the approach adopted in this book. However, as Equations (7.1), (7.3), (7.4) and (7.5) show, when used with cyclic processes, the sign convention results in negative signs appearing in some equations and it can make arguments difficult to follow. In such circumstances it is often convenient to treat all the Qs and Ws as numerical quantities, symbols $|Q|$, and so on, and to indicate the direction of energy transfer either on a diagram or in words. For readers who prefer this approach, the discussion of Carnot's theorem and its corollary, which follows in Section 7.4, is developed, using numerical quantities, in the appendix to this chapter. Note, however, that the discussion in Chapter 8 is not easily developed except by strict adherence to the original sign convention.

7.4 Carnot's theorem and its corollary

Carnot's theorem states that: no engine operating between two reservoirs can be more efficient than a Carnot engine operating between the same reservoirs.

Since the Carnot engine operates reversibly, all other types of engine must be irreversible. The theorem may be proved as follows.

Consider two heat engines, denoted by I and R, operating between heat reservoirs at temperatures θ_1 and θ_2, respectively, with $\theta_1 > \theta_2$. Engine R is a reversible (Carnot) engine and engine I is irreversible. When engine I operates in the forward sense, let it absorb heat Q_1 per cycle from the reservoir at temperature θ_1 and heat Q_2 per cycle from that at temperature θ_2, and have work W done on it per cycle. The corresponding quantities for engine R are Q_1', Q_2' and W' (see Figure 7.3).

Assume that the rates of working of the two engines are such that, when engine I completes m cycles, engine R completes n cycles. The amounts of working substance in the two engines may be chosen at will and are arranged so that $m|Q_1| = n|Q_1'|$. Now let the two engines be coupled together so that engine I drives engine R backwards. This makes Q_1 and Q_1' have oppositve signs but, because R is reversible, does not change the

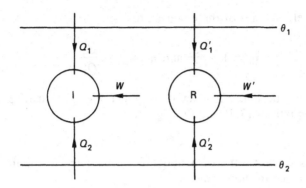

Figure 7.3 *Arrangement to demonstrate Carnot's theorem*

magnitude of the heat it absorbs per cycle from the reservoir at temperature θ_1. In one complete cycle of this composite engine, engine I performs m of its cycles and engine R performs n of its cycles. In each cycle of the composite engine the heat absorbed from the reservoir at temperature θ_1 has been arranged to be zero, while the heat absorbed from the reservoir at temperature θ_2 is $mQ_2 + nQ_2'$. Applying the first law to one cycle of the complete device gives

$$mQ_2 + nQ_2' + W_c = 0 \tag{7.6}$$

where W_c is the total work done on the working substances. It can be seen that the first law allows three possibilities for the behaviour of the composite engine: (1) $mQ_2 + nQ_2'$ is positive and W_c is negative; (2) $mQ_2 + nQ_2'$ and W_c are both zero; (3) $mQ_2 + nQ_2'$ is negative and W_c is positive.

If the Kelvin statement of the second law is not to be violated, $mQ_2 + nQ_2'$ cannot be positive, as this would imply that the composite engine is a cyclic device that extracts a quantity of heat $mQ_2 + nQ_2'$ from its surroundings and performs an equal amount of work on its surroundings, while producing no other effect. Therefore,

$$mQ_2 + nQ_2' \leqslant 0$$

The amounts of working substance have been chosen so that

$$mQ_1 + nQ_1' = 0$$

i.e.

$$mQ_1 = -nQ_1'$$

Therefore,

$$\frac{mQ_2}{mQ_1} - \frac{nQ_2'}{nQ_1'} \leqslant 0 \tag{7.7}$$

The thermal efficiencies of the engines are

$$\eta_I = 1 + \frac{Q_2}{Q_1} \text{ and } \eta_R = 1 + \frac{Q_2'}{Q_1'} \tag{7.8}$$

where η_I is the thermal efficiency of engine I and η_R that of engine R. Combining equations (7.7) and (7.8) gives

$$\eta_I \leqslant \eta_R \tag{7.9}$$

This result is Carnot's theorem and holds whatever the nature of the working substance in the engine.

There is an important corollary to this theorem which can be obtained by replacing the engine I by a second reversible engine R'. Let this engine absorb a quantity of heat Q_1 per cycle from the reservoir at temperature θ_1 and Q_2 from the reservoir at temperature θ_2. Again assume that R' completes m cycles while R completes n cycles and arrange the amounts of working substance so that $m|Q_1| = n|Q_1'|$. These two reversible engines are then coupled together to form a composite engine, the senses of the individual cycles being such that Q_1 and Q_1' are of opposite sign. In one complete cycle of the composite engine there is no change in the internal energy of the working substances; the total heat absorbed from the reservoir at temperature θ_1 is $mQ_1 + nQ_1'$, which is zero; and that absorbed from the reservoir at temperature θ_2 is $mQ_2 + nQ_2'$. To avoid violation of the Kelvin statement of the second law, $mQ_2 + nQ_2'$ cannot be positive and, therefore,

$$mQ_2 + nQ_2' \leqslant 0 \tag{7.10}$$

Since both engines are thermodynamically reversible, reversing the senses of their cycles simply changes the signs of the interactions. Therefore, when the composite engine is operated in reverse,

$$-mQ_2 - nQ_2' \leqslant 0 \tag{7.11}$$

Conditions (7.10) and (7.11) must both be satisfied simultaneously for reversible engines and this is only possible if

$$mQ_2 + nQ_2' = 0$$

The amounts of working substance have been chosen so that

$$mQ_1 + nQ_1' = 0$$

and, therefore,

$$\frac{Q_1}{Q_2} = \frac{Q_1'}{Q_2'} \tag{7.12}$$

The efficiencies are given by

$$\eta_R = 1 + Q_2'/Q_1' \text{ and } \eta_{R'} = 1 + Q_2/Q_1$$

so that

$$\eta_R = \eta_{R'} \tag{7.13}$$

This result is the corollary to Carnot's theorem: all reversible engines operating between the same two reservoirs have equal efficiencies. No conditions were placed on the natures of the working substances used in the Carnot engines. From this it follows that the efficiency of a reversible engine operating between two reservoirs is independent of the nature of the working substance; it can be a function of the temperatures of the reservoirs only.

In summary, for any reversible engine R operating between reservoirs at temperatures θ_1 and θ_2,

$$\eta_R = -\frac{W}{Q_1} = \frac{Q_1 + Q_2}{Q_1} = 1 + \frac{Q_2}{Q_1} = f(\theta_1, \theta_2) \tag{7.14}$$

where f is a universal function of θ_1 and θ_2. Further, η_R cannot be exceeded by any irreversible engine operating between the same reservoirs.

7.5 Universal temperature functions

The existence of a property — the thermal efficiency of a Carnot engine — that depends only on temperatures and not on the properties of any particular substance or class of substances allows some of the arbitrariness to be removed from the process of determining temperatures. For a Carnot engine that absorbs heat Q_1 per cycle from a reservoir at a temperature θ_1 and heat Q_2 from a reservoir at temperature θ_2, Equation (7.14) may be written

$$-\frac{Q_1}{Q_2} = \phi(\theta_1, \theta_2) \tag{7.15}$$

where ϕ is another universal function of θ_1 and θ_2. The negative sign is used because Q_1 and Q_2 must be opposite sign if the Kelvin statement of the second law is not to be violated. As it stands, equation (7.15) is not a basis for defining temperatures, but it may readily be shown that $\phi(\theta_1, \theta_2)$ is separable into the form $F(\theta_1)/F(\theta_2)$, where F is also a universal function. With $\phi(\theta_1, \theta_2)$ expressed in this form, Equation (7.15) can be used to define temperatures. A proof that Q is a function of the temperature of the reservoirs only is as follows.

For a given mass of a particular working substance in a Carnot engine, choose any two reversible adiabatic processes, represented by AE and BD in the graph of generalised force X against generalised displacement x (Figure 7.4). The equilibrium states of the adiabatic processes are linked by three reversible isothermal processes, represented by AB, FC and ED, corresponding to empirical temperatures θ_1, θ_2 and θ_3, respectively. If the amounts of heat absorbed by the working substance when the isotherms are traversed in the direction of increasing x are Q_1, Q_2 and Q_3, respectively, then from, Equation (7.15),

$$Q_1/Q_2 = \phi(\theta_1, \theta_2)$$
$$Q_2/Q_3 = \phi(\theta_2, \theta_3)$$

and

$$Q_1/Q_3 = \phi(\theta_1, \theta_3)$$

Therefore,

$$\phi(\theta_1, \theta_2) = \phi(\theta_1, \theta_3)/\phi(\theta_2, \theta_3) \qquad (7.16)$$

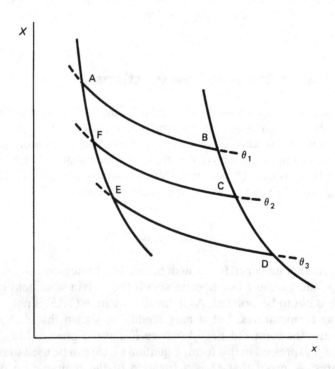

Figure 7.4 *Reversible isotherms and reversible adiabatics for the working substance of a Carnot engine*

Since θ_3 is an independent variable which appears only on the right-hand side of Equation (7.16), it must vanish from the equation, which means that $\phi(\theta_1,\theta_2)$ must be of the form $F(\theta_1)/F(\theta_2)$. Therefore, for a Carnot engine operating between reservoirs at temperatures θ_1 and θ_2, Equation (7.15) becomes

$$-\frac{Q_1}{Q_2} = \frac{F(\theta_1)}{F(\theta_2)} \tag{7.17}$$

In principle, $F(\theta)$ can be determined, within the limits of experimental error, using any chosen empirical scale of temperature, by making measurements on engines that are, as closely as possible, performing Carnot cycles. However, it is also possible to use Equation (7.17) to define a new temperature by choosing a function for $F(\theta)$. Each of these functions of θ, which may be called *universal temperature functions,* defines a property called a *universal temperature.* The only conditions to be satisfied by the choice for $F(\theta)$ are that the isotherms be uniquely numbered and that the hotness convention be followed. These restrictions still allow an indefinite number of choices for $F(\theta)$, so that there is an indefinite number of universal temperature functions.

In making a choice of a universal temperature function to be adopted as standard, it is convenient to choose one on which the characteristic properties of the Carnot engine are expressed in a simple form. Two possibilities, first suggested by Lord Kelvin, will now be considered.

7.6 Efficiency and work temperature functions

In 1848 Lord Kelvin suggested that the universal temperature function chosen as standard be given by

$$-\frac{Q_1}{Q_2} = \frac{e^{\tau_1}}{e^{\tau_2}} \tag{7.18}$$

This equation defines a universal temperature, denoted by τ, in which the heat Q absorbed per cycle by a Carnot engine from a reservoir at a temperature τ is proportional to e^{τ}. Equation (7.18) actually gives a ratio of temperatures. To define a unit of temperature, a value τ_0 must be assigned to some arbitrarily chosen fixed temperature. If a Carnot engine is used as a thermometer and absorbs heat Q_0 per cycle from a reservoir at temperature τ_0 and heat Q from a reservoir at temperature τ, the equation defining τ is

$$e^{\tau} = -e^{\tau_0} \frac{Q}{Q_0} \tag{7.19}$$

or

$$\tau = \tau_0 + \ln(-Q/Q_0) \tag{7.20}$$

Q and Q_0 must, of course, be of opposite sign, so that the lowest temperature attainable is that for which Q is zero. This corresponds to a value for τ of minus infinity, whatever the size of unit used.

Using this temperature function, if two Carnot engines operate between the same reversible adiabatics, their efficiencies are equal when the difference between the intake and output temperatures is the same for each (see Exercise 1). For this reason the function is often referred to as the *efficiency universal temperature function*.

The efficiency function received little support and in 1854 Lord Kelvin made a second suggestion — namely, that the universal temperature function adopted should define a temperature T by the equation

$$-\frac{Q_1}{Q_2} = \frac{T_1}{T_2} \tag{7.21}$$

Like Equation (7.18), this equation gives a ratio between two temperatures, so that Equation (7.21) defines T except for an arbitrary constant of proportionality which gives the size of the unit and is obtained by assigning the value T_0 to an arbitrarily chosen fixed temperature. Then, if the quantities of heat absorbed per cycle by a Carnot engine operating between reservoirs at temperatures T and T_0 are Q and Q_0, respectively, T is defined by

$$T = -T_0 \frac{Q}{Q_0} \tag{7.22}$$

Since Q and Q_0 must be of opposite sign, T can never be negative. The smallest possible value for $|Q|$ is zero and the corresponding value of T is zero. This temperature is referred to as the 'absolute zero of temperature' and, with the function chosen by Equation (7.21), is a finite number of units below the temperature T_0.

Using this temperature function, it is readily shown that, if two Carnot engines operate between the same reversible adiabatics, they do equal amounts of work per cycle when the differences between the intake and output temperatures are equal (see Exercise 3). Therefore, the function is referred to as the *work universal temperature function*.

All universal temperature functions are related. For the two functions considered, combining Equations (7.19) and (7.22) gives

$$\frac{T}{T_0} = \frac{e^\tau}{e^{\tau_0}}$$

which may be written

$$T = A \, e^\tau \tag{7.23}$$

where $A = T_0/e^{\tau_0} = $ constant.

7.7 The thermodynamic temperature function

The work universal temperature function has been adopted as the international standard and is called the *thermodynamic temperature function*. Temperatures defined by this function are *thermodynamic temperatures*, symbol T. The unit of thermodynamic temperature is the kelvin, symbol K, which is the fraction 1/273.16 of the temperature of the triple point of water (see Section 13.1). When a Carnot engine operates between a reservoir at a temperature T, measured in kelvins, and one at a temperature of 273.16 K, the thermodynamic temperature T is given by

$$T/K = -273.13 \, \frac{Q}{Q_3} \tag{7.24}$$

where Q is the heat absorbed per cycle at the temperature T and Q_3 that absorbed at the triple point of water.

The triple point of water is that state of the water substance in which ice, liquid water and water vapour coexist in equilibrium. It is chosen as the standard fixed point because of its high reproducibility (see Section 14.5). The triple point of water may be realised using the cell shown in Figure 7.5. Pure water is introduced into a vessel which is then pumped through P to remove the air before sealing off. The vessel then contains only pure water and its vapour. A freezing mixture is then placed in the re-entrant cavity C to produce a thick layer of ice I around this cavity. When the freezing mixture is replaced by the sensing element of a thermometer, a thin layer of ice in contact with the cavity is melted. The triple-point temperature is obtained wherever the ice is in equilibrium with a liquid–vapour interface. A simple correction gives the variation of temperature with depth below the liquid–vapour surface. Since the temperature of the triple point of water is only about 0.01 K above the normal melting temperature of ice, the cell is kept in a bath of melting ice to reduce heat transfer from the surroundings.

7.8 Absolute zero

Equation (7.24) shows that, for a Carnot engine operating between two chosen reversible adiabatic paths, the heat Q absorbed in each reversible

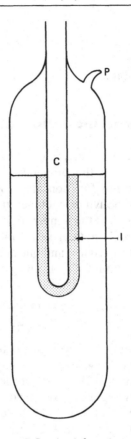

Figure 7.5   *A triple-point cell*

isothermal process is proportional to the thermodynamic temperature T of that process. The smallest value that Q can take is zero and the corresponding value of T is zero, known as *absolute zero*. Absolute zero is, therefore, the temperature at which there is no heat transfer in a reversible, isothermal process. This definition holds for all systems and is an operational definition in terms of large-scale properties. Nothing is ascribed to the properties and behaviour of the constituent molecules, atoms or ions.

Now consider a Carnot refrigerator operating between a reservoir at a temperature T_0 and a system of finite mass at a temperature T. In one cycle, the work W needed to extract a quantity of heat dQ from the finite mass is

$$W = -dQ - dQ_0$$

where dQ_0 is the heat absorbed from the reservoir at temperature T_0. Since

$$\frac{\mathrm{d}Q}{\mathrm{d}Q_0} = -\frac{T}{T_0}$$

$$W = \mathrm{d}Q\left(\frac{T_0 - T}{T}\right) = \mathrm{d}Q\left(\frac{T_0}{T} - 1\right) \tag{7.25}$$

As T tends to zero, so W tends to infinity, showing that absolute zero is a limiting temperature that can be approached as closely as is required by the expenditure of sufficient work, and in this sense constitutes a well-defined level of temperature. Note, however, that the value of the lowest available temperature at a finite number of units below the triple point temperature of water is a property of the particular temperature function chosen and not an intrinsic property of temperature.

7.9 Celsius temperature

The *Celsius temperature*, symbol t, has as its unit the degree Celsius, symbol °C, which is equal to the kelvin. The *Celsius temperature scale* is defined by the equation

$$t/°C = T/K - 273.15 \tag{7.26}$$

t has no independent definition. Differences in temperature may be expressed in kelvins or °C, but the former is preferred.

7.10 The measurement of thermodynamic temperature

Thermodynamic temperatures are defined in terms of the behaviour of a Carnot engine, the defining equation being Equation (7.24). Unfortunately, the Carnot engine is an ideal device and cannot be realised exactly in practice, and so it would appear that precise determinations of thermodynamic temperature are impossible. However, it will be shown later (Sections 9.3 and 9.4) that this difficulty can be overcome by making appropriate measurements on real gases. Until that stage is reached, temperatures must be determined by the performance of Carnot cycles under conditions as near to ideal as possible, determining Q/Q_3 and using Equation (7.24).

■ EXAMPLE

Q. A certain Carnot engine absorbs 1000 J of heat per cycle from a reservoir at 400 K and heat Q per cycle from a reservoir at 300 K. Calculate the value of Q, the efficiency η of the engine and the work W done on the working substance per cycle.

A. If Q_1 is the heat absorbed per cycle from the high-temperature reservoir, the efficiency is given by

$$\eta = -W/Q_1$$

and the work done on the working substance per cycle is

$$W = -Q_1 - Q$$

From the definition of thermodynamic temperature, $Q_1/Q = -(400)/(300)$. Now, using the sign convention, $Q_1 = +1000$ J. Therefore,

$$Q = -Q_1 \cdot \frac{300}{400} = -(+1000) \cdot \frac{300}{400} = -750 \text{ J}$$

The minus sign indicates that this heat is rejected to the low-temperature reservoir. This value of Q may now be used to calculate W:

$$W = -Q_1 - Q = -(+1000) - (-750)$$
$$= -1000 + 750$$
$$= -250 \text{ J}$$

Here the minus sign indicates that work is done on the surroundings. The efficiency is then given by

$$\eta = -\frac{W}{Q_1} = -\frac{(-250)}{1000} = +0.25 \text{ or } 25\%$$

It should be noted that, even for an ideal engine, the efficiency is not 100%.

Appendix: An alternative approach to cyclic processes

In this approach all amounts of heat and work are treated as positive quantities and denoted by $|Q|$ and $|W|$, respectively. A heat engine operating between two heat reservoirs at temperatures θ_1 and θ_2 ($\theta_1 > \theta_2$) is then represented as in Figure 7.6, in which the arrows denote the actual direction in which the energy transfers occur. In one cycle the working

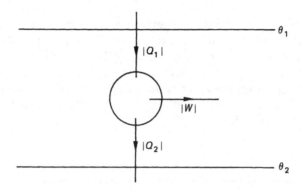

Figure 7.6 *Symbolic representation of a heat engine operating between two reservoirs*

substance extracts heat $|Q_1|$ from the high-temperature reservoir and rejects heat $|Q_2|$ to the low-temperature reservoir, while doing work $|W|$ on the surroundings. There is no change in the internal energy of the working substance after one complete cycle, so the first law for one cycle is

$$|Q_1| - |Q_2| = |W| \tag{7.27}$$

The thermal efficiency η of the engine is given by

$$\eta = \frac{|W|}{|Q_1|} \tag{7.28}$$

Let two engines I and R operate between these two reservoirs, and adjust the amounts of working substance in each so that, when operating in the forward sense, each does the same amount of work $|W|$ per cycle on the surroundings. When engine I operates in the forward sense, it extracts heat $|Q_1|$ from the high-temperature reservoir and rejects heat $|Q_2|$ to the low-temperature reservoir in each cycle. The corresponding quantities for engine R are $|Q_1'|$ and $|Q_2'|$, respectively.

The efficiency of engine I is η_I, where

$$\eta_I = \frac{|W|}{|Q_1|} = \frac{|Q_1| - |Q_2|}{|Q_1|} \tag{7.29}$$

while that of engine R is η_R, where

$$\eta_R = \frac{|W|}{|Q_1'|} = \frac{|Q_1'| - |Q_2'|}{|Q_1'|} \tag{7.30}$$

Now couple engines I and R together so that I drives R backwards. Since engine R is reversible, in each cycle it now extracts heat $|Q_2'|$ from the low-temperature reservoir and rejects heat $|Q_1'|$ to the high-temperature reservoir. The two engines coupled in this way constitute a self-acting device, since engine I supplies exactly the amount of work per cycle needed to drive R backwards. The heat and work transfers for this coupled engine are as shown in Figure 7.7.

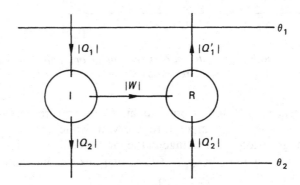

Figure 7.7 *Engine I operating a Carnot engine R in reverse*

Assume that engine I is more efficient than engine R. Then, from Equations (7.29) and (7.30),

$$\frac{|Q_1| - |Q_2|}{|Q_1|} > \frac{|Q_1'| - |Q_2'|}{|Q_1'|} \tag{7.31}$$

But, from the first law,

$$|Q_1| - |Q_2| = |W| = |Q_1'| - |Q_2'|$$

so that

$$\frac{|W|}{|Q_1|} > \frac{|W|}{|Q_1'|}$$

or

$$|Q_1'| > |Q_1| \tag{7.32}$$

For this coupled engine the net heat extracted from the low-temperature reservoir is $|Q_2'| - |Q_2|$, which, from the first law is equal to $|Q_1'| - |Q_1|$. Relation (7.32) shows that this is a positive quantity. The engine also delivers to the high-temperature reservoir a net quantity of heat given by $|Q_1'| - |Q_1|$. Therefore, this self-acting device produces no effect other than the transfer of quantity of a heat from a cold reservoir to a hot reservoir. This is a violation of the Clausius statement of the second law, so that the original assumption, that $\eta_I > \eta_R$, must be wrong. Therefore,

$$\eta_I \leqslant \eta_R \tag{7.33}$$

which is Carnot's theorem.

The corollary to Carnot's theorem is proved by replacing engine I with a second reversible engine, R'. Then, if R drives R' backwards, Carnot's theorem states that

$$\eta_R \leqslant \eta_{R'}$$

while, if R' drives R backwards,

$$\eta_{R'} \leqslant \eta_R$$

Therefore,

$$\eta_R = \eta_{R'}$$

which is the required result.

■ EXERCISES

1 Two Carnot engines operate between the same two reversible adiabatics. Engine A absorbs heat from reservoirs at temperatures τ_1 and τ_2 (defined by the efficiency temperature function) and engine B absorbs heat from reservoirs at temperatures τ_3 and τ_4. Show that, if $|\tau_1 - \tau_2| = |\tau_3 - \tau_4|$, the efficiencies of engines A and B are equal.

2 If the temperature of the triple point of water has the value zero thomsons, symbol Th, when temperature is defined by the efficiency temperature function, calculate the temperature in thomsons corresponding to 100 K.

3 Two Carnot engines operate between the same two reversible adiabatics. Engine A absorbs heat from reservoirs at temperatures T_1 and T_2, while engine B absorbs heat from reservoirs at temperatures T_3 and T_4. Show that, if $|T_1 - T_2| = |T_3 - T_4|$, the work done on the working substance per cycle is the same for both engines.

4 Show that, for a Carnot engine operating between reservoirs at temperatures T_1 and T_2 ($T_1 > T_2$), the thermal efficiency is given by

$$\eta_R = \frac{T_1 - T_2}{T_1}$$

5 A Carnot engine operates between reservoirs at temperatures T_1 and T_2 ($T_1 > T_2$). Is it more effective to increase the efficiency of the engine

(a) by increasing T_1 by a certain amount, keeping T_2 constant;
(b) by decreasing T_2 by the same amount, keeping T_1 constant?

6 A student applies to a research trust for money to develop an engine that is said to extract 5000 J of heat per cycle from a reservoir at 400 K, reject 3500 J of heat per cycle to a reservoir at 300 K and do 1500 J of work per cycle on the surroundings. Is the application likely to be successful?

7 An inventor claims to have designed an engine that does 1200 J of work on the surroundings per cycle while receiving 1000 J of heat per cycle from a single reservoir, Does the engine violate

(a) the first law,
(b) the second law,
(c) both first and second laws,
(d) neither the first nor the second law?

8 A lecturer claims to have constructed a device that rejects 100 J of heat to a single reservoir while absorbing 100 J of work in each cycle. Does this cyclic device violate

(a) the first law,
(b) the second law,
(c) both first and second laws,
(d) neither the first nor the second law?

9 Derive expressions for the coefficient of performance of a Carnot refrigerator and of a Carnot heat pump operating between reservoirs at temperatures T_1 and T_2 ($T_1 > T_2$). Plot a graph of C.o.P. against T_2/T_1 for each device.

10 A piece of rubber under tension has its equilibrium states specified by the coordinates tension F, length L and temperature T. When L is allowed to decrease isothermally, there is a flow of heat from the surroundings to the rubber, and vice versa. If a rubber band is used as the working substance of a Carnot engine operating between reservoirs at temperatures T_1 and T_2 $(T_1 > T_2)$, draw a graph in F–L space representing the cycle of operations, assuming that the material obeys Hooke's law. Note that the linear expansivity of rubber is negative.

Notes

1 The general concept of a cyclic process does not require that the process be quasistatic, simply that the final condition of the system be identical with the initial condition. However, for a classical thermodynamic statement, a cyclic process must start and end with the system in the same equilibrium state.

2 Carnot pointed out in 1824 that, to obtain the maximum amount of work per cycle from an engine operating between two reservoirs at different temperatures, two conditions must be satisfied : (1) There must be no thermal contact between bodies whose temperatures differ by more than an infinitesimal amount, since the direct thermal contact of bodies whose temperatures do differ by a finite amount means that some of the potentiality of the temperature difference to produce work has been wasted. (2) There must be no mechanical losses such as friction. As will be shown in Section 7.4, under these conditions there is thermodynamic equilibrium at every stage of the process and an absence of dissipative effects, so that the cycle is thermodynamically reversible.

3 A proof that the path for $Q = 0$ is unique for a system with only two degrees of freedom is as follows.

Consider a system with independent coordinates X and x, where X is the generalised force and x is the generalised displacement. For an infinitesimal reversible process,

$$dQ = dU - dW = dU - Xdx$$

Now U is a state function and may, therefore, be written

$$U = U(X, x)$$

Then

$$dU = \left(\frac{\partial U}{\partial X}\right)_x dX + \left(\frac{\partial U}{\partial x}\right)_X dx$$

Therefore,

$$\mathrm{d}Q = \left(\frac{\partial U}{\partial X}\right)_x \mathrm{d}X + \left[\left(\frac{\partial U}{\partial x}\right)_X - X\right] \mathrm{d}x$$

and in an adiabatic process $\mathrm{d}Q$ is zero. The coefficients of $\mathrm{d}X$ and $\mathrm{d}x$ are functions of state, so that the equation describing the reversible adiabatic process may be written

$$F_1(X, x) \, \mathrm{d}X + F_2(X, x) \, \mathrm{d}x = 0$$

which may be integrated to give a uniquely defined path for the process.

8

Entropy

Thermodynamic temperature T is so defined that, for a Carnot engine that absorbs heat Q_1 per cycle from a reservoir at a temperature T_1 and heat Q_2 per cycle from a reservoir at a temperature T_2,

$$-\frac{Q_1}{Q_2} = \frac{T_1}{T_2} \tag{8.1}$$

This equation may be written

$$\frac{Q_1}{T_2} + \frac{Q_2}{T_2} = 0$$

or even, with what might appear as a rather excessive sophistication,

$$\sum \frac{Q}{T} = 0 \tag{8.1a}$$

An irreversible engine operating in a cycle between the same reservoirs cannot be more efficient than a Carnot engine. Therefore, if an irreversible engine absorbs heats Q_1' and Q_2' per cycle from the reservoirs at temperatures T_1 and T_2, respectively,

$$\frac{Q_2}{Q_1} \geqslant \frac{Q_2'}{Q_1'} \quad \text{(see Equation 7.7)}$$

Substituting for Q_2/Q_1 from Equation (8.1) gives

$$-\frac{T_2}{T_1} \geqslant \frac{Q_2'}{Q_1'}$$

or

$$\frac{Q_2'}{T_2} \leqslant - \frac{Q_1'}{T_1}$$

or

$$\sum \frac{Q}{T} \leqslant 0 \qquad\qquad (8.2)$$

It might now be asked whether there is a result analogous to Equation (8.2) for a general cyclic process, and it is the answer to this question that leads to the concept of entropy.

8.1 The inequality of Clausius

To investigate the question posed above, consider an arbitrarily chosen closed system A and let it perform a cyclic process which may be reversible or irreversible. The external effects of this cyclic process are the performance of work and the absorption of heat. Assume that the heat exchange takes place between the system and a single heat reservoir at a temperature T_0.

The cyclic process executed by this system may be divided into a succession of elementary steps. In each of these steps the system may not have a single definable temperature, but a small region of its boundary can always be found where the temperature is sensibly uniform. It is then possible to make the heat exchange take place in a reversible manner by making use of a Carnot engine C operating in infinitesimal cycles. The cycle of the Carnot engine must be adjusted so that one infinitesimal reversible isothermal process occurs at the temperature T of the identified small area of the boundary and the other at the temperature T_0 of the reservoir (see Figure 8.1). In this way the heat transfer can be made reversible, even when the process that the system is performing is irreversible.

Each element of the cyclic process performed by the system A is then achieved in the following steps.

1   The Carnot engine C is in equilibrium in thermal contact with the reservoir at a temperature T_0. The system A is in its initial state.
2   C is brought by a reversible adiabatic process to a temperature T and is placed in thermal contact with that part of the surface of A that is also at a temperature T.

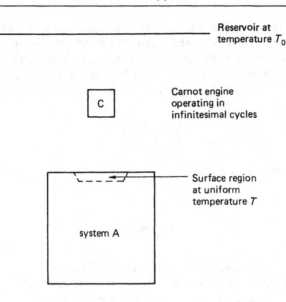

Figure 8.1 *Arrangement to demonstrate the inequality of Clausius*

3 C undergoes an infinitesimal reversible isothermal process at a temperature T, during which the system A absorbs a quantity of heat q from C.

4 C undergoes a reversible adiabatic process that returns its temperature to T_0.

5 It is placed in thermal contact with the reservoir and undergoes a reversible isothermal process at a temperature T_0 until it returns exactly to its initial state. From the definition of thermodynamic temperature, C absorbs a quantity of heat $q\,T_0/T$ during this process.

During one complete cycle performed by the system A, the heat Q absorbed by C from the reservoir is given by

$$Q = \oint \frac{q\,T_0}{T} \tag{8.3}$$

where the integral is taken round the cycle of A. Assume that, when the system A completes one cycle, the Carnot engine C completes a whole number of its cycles. Then there is no change in the internal energy of either A or C when A has completed one cycle and

$$Q + W = 0 \tag{8.4}$$

where W is the work done on both A and C in one cycle of A.

If Q were positive, Equation (8.4) would indicate that a cyclic device exists that produces no effect other than the extraction of a certain quantity of heat from a reservoir and the performance of an equal amount of work on the surroundings. This would violate Kelvin's statement of the second law. It must be concluded, therefore, that for any cyclic process

$$\oint \frac{q\, T_0}{T} \leq 0$$

Since T_0 is constant and necessarily positive, the above result may be written

$$\oint \frac{q}{T} \leq 0 \qquad (8.5)$$

Statement (8.5) is known as the inequality of Clausius.

If the process performed by A is, in fact, reversible, it can be performed in the opposite sense, with q for each element of the original cycle becoming $-q$. A repetition of the above reasoning then gives

$$-\oint \frac{q}{T} \leq 0 \qquad (8.6)$$

For a reversible cyclic process conditions (8.5) and (8.6) must be satisfied simultaneously, and this is only possible if

$$\oint \frac{q}{T} = 0 \qquad (8.7)$$

Equation (8.7) is the theorem of Clausius and is valid for any reversible cyclic process. Therefore, in relation to Equation (8.5), the inequality holds for any irreversible cyclic process and the equality for any reversible cyclic process. Relation (8.5), therefore, contains the answer to the question raised at the end of the introduction to this chapter.

It should be noted that the temperature T is strictly that of the source that supplies the heat to the system A. In a reversible process T is also the temperature of the system, but in an irreversible process the system need not have a unique temperature.

8.2 Entropy

Consider a system performing a reversible cyclic process. When two independent coordinates X and x are sufficient to specify equilibrium states of the system, the path of the process may be represented by a closed curve

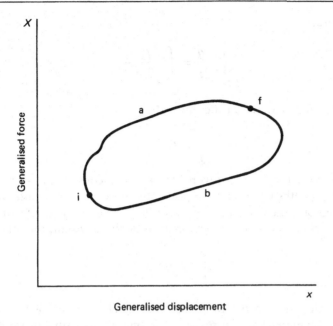

Generalised force

Generalised displacement

Figure 8.2 *To show that entropy is a function of the state of a system*

in X–x–T space, and this may be projected onto the X–x plane to give another closed curve, as in Figure 8.2. However, the following argument applies to systems of any complexity.

Let i and f be any two arbitrarily chosen (equilibrium) states of the reversible cyclic process. The theorem of Clausius states that, for any reversible cyclic process,

$$\oint \frac{q}{T} = 0$$

where q is the heat absorbed by the system in the element of the process that takes place at temperature T. This cycle may be considered to be performed in two parts — namely, iaf and fbi — and the above equation may then be written

$$\oint \frac{q}{T} = \int_{iaf} \frac{q}{T} + \int_{fbi} \frac{q}{T} = 0 \tag{8.8}$$

Since the cyclic process is reversible,

$$\int_{fbi} \frac{q}{T} = -\int_{ibf} \frac{q}{T}$$

and, therefore,

$$\int_{iaf} \frac{q}{T} = \int_{ibf} \frac{q}{T} \qquad (8.9)$$

Equation (8.9) shows that the integral

$$\int q/T$$

has a single value for the two reversible processes considered — that is, those following paths iaf and ibf, respectively. Since these two paths were chosen in an arbitrary manner, this result is generally true: for all reversible processes between states i and f of a given system, the integral

$$\int_i^f q_R/T$$

has a single value, irrespective of the path taken. The subscript R has been inserted as a reminder that the result refers only to reversible processes.

The existence of a path-independent quantity

$$\int_i^f q_R/T$$

implies the existence of a property of the system such that, in going from an initial equilibrium state i to a final equilibrium state f by any reversible process, the change in the value of the property is given by

$$\int_i^f q_R/T$$

This property of the system is called the *entropy* of the system, symbol S, so that

$$\Delta S = S_f - S_i = \int_i^f q_R/T = \int_i^f dQ_R/T \qquad (8.10)$$

The unit of entropy is JK^{-1}. In an infinitesimal reversible process

$$dS = dQ_R/T \qquad (8.11)$$

It should be recalled that alternative terms for '(thermodynamic) property' are 'state function' and '(thermodynamic) coordinate'. Both of these will be applied to entropy.

Since i and f are equilibrium states of the system, S is defined for equilibrium states only. Each equilibrium state of a system is associated with a definite value of the entropy (relative to that in a chosen reference state, since Equation (8.10) involves ΔS only), no matter by what reversible path that state is achieved. It will be noticed that there is a strong similarity between Equation (8.10) and Equation (4.1), which defined internal energy.

Further, since entropy is a state function or coordinate, the change in entropy of a given system must have a single value for a change from an initial equilibrium state i to a final equilibrium state f, whatever the nature of the process. When the process is reversible, $S_f - S_i$ is given by

$$\int_i^f q_R/T$$

but when the process is irreversible, the change in entropy of the system is the same but is no longer given by Equation (8.10). This does not mean that there is anything wrong with the concept of entropy but simply that Equation (8.10) can only be applied to reversible processes. To calculate the entropy change in a system that goes from equilibrium state i to equilibrium state f by means of an irreversible process, it is necessary to devise a reversible process linking states i and f and calculate

$$\int_i^f q_R/T$$

for it. Since the entropy change is the same for all reversible processes between states i and f, any convenient reversible process will do for this calculation. This does not imply that reversible and irreversible processes produce the same effects in the universe as a whole. For given initial and final equilibrium states, the change in entropy of the system is the same whatever the process, but the change in entropy of the surroundings will depend on the nature of the process.

As an example, consider n mol of molecules of an ideal gas undergoing a free expansion from an equilibrium state i, characterised by pressure p_i, volume V_i and temperature T_i, to a final equilibrium state f where the respective values are p_f, V_f, T_f. As was shown in Section 6.2, in any free expansion the change in the internal energy U of the system is zero. Further, for an ideal gas, a free expansion produces no change in temperature. Therefore, an equivalent reversible process linking states i and f is a reversible isothermal expansion at a temperature T_i $(=T_f)$. Applying the first law to this equivalent process gives

$$\Delta U = 0 = Q + W$$

where Q is the heat absorbed by the gas and W is the work done on the gas.

Now

$$W = - \int_i^f p\,dV$$

and the equation of state is Boyle's law — that is, $p\,\dfrac{V}{n} = f(T)$. Therefore,

$$W = - \int nf(T)\,\frac{dV}{V}$$

$$= -nf(T)\,\ln(V_f/V_i)$$

and $\left. \begin{array}{c} \\ \\ Q = -W = nf(T)\,\ln(V_f/V_i) \end{array} \right\}$ at constant T

The change in entropy of the gas is, therefore,

$$\Delta S = S_f - S_i = \frac{Q}{T_i} = \frac{nf(T)\,\ln(V_f/V_i)}{T_i}$$

But

$$f(T) = p_iV_i/n = p_fV_f/n$$

so that

$$\Delta S = \frac{p_iV_i\,\ln\,(V_f/V_i)}{T_i} \qquad (8.12)$$

ΔS is positive: the entropy of the gas has increased. Note that for the real process — the free expansion — the entropy change of the surroundings is zero, so that the entropy change of the universe is positive. However, in the equivalent reversible process used to calculate ΔS, the entropy of the system increases by Q/T_i but that of the surroundings, which could be a heat reservoir at temperature T_i, decreases by Q/T_i so that the entropy change of the universe is zero.

Since entropy is a thermodynamic coordinate, it can be used in representations of equilibrium states. For example, the equilibrium states of a fixed mass of gas can be represented by a surface in p–V–S space, rather than p–V–T space. A representation that is generally applicable to systems with two degrees of freedom, and has useful properties, is the representation in U–V–S space, the so-called Gibbs surface (see Exercise 5).

In an adiabatic reversible process Q is zero and, therefore, ΔS is zero. Consequently, a reversible adiabatic process is represented by a line drawn with S equal to a constant on the appropriate surface. The equation for the behaviour under reversible adiabatic conditions is obtained from the equation of state, subject to the condition that S is constant. Such a process is termed *isentropic*. Since S is a function of state, it follows that the reversible adiabatic through a given equilibrium state of a system is uniquely defined, whatever the complexity of the system. This resolves the uniqueness problem that was put to one side in Section 7.3.

8.3 Entropy and work

Consider a Carnot engine operating between reservoirs at temperatures T and T_0. In one cycle the work W done on the working substance is

$$W = -(Q + Q_0) \tag{8.13}$$

where Q and Q_0 are the quantities of heat absorbed per cycle by the working substance at temperatures T and T_0, respectively. From the definition of thermodynamic temperature,

$$\frac{Q}{Q_0} = -\frac{T}{T_0}$$

so that

$$W = -\left(Q - \frac{Q T_0}{T}\right)$$

$$= -\frac{Q}{T}(T - T_0) = -\Delta S(T - T_0) \tag{8.14}$$

where ΔS is the increase in entropy of the working substance (the system) by virtue of the heat absorbed at the temperature T. In terms of Equation (8.14), the unit of entropy is that which is capable of doing unit amount of work for unit drop in temperature in a Carnot cycle. Alternatively, if the entropy change of a system in a reversible isothermal process at a temperature T is ΔS, the maximum amount of work that can be obtained from the energy transferred is obtained by means of a Carnot engine (since that is the most efficient device) and is of magnitude $\Delta S(T - T_0)$, where T_0 is the lowest temperature available.

In the light of Equation (8.14), entropy may be said to possess the capacity to do work by virtue of its elevation of temperature above that of the lowest temperature available. The amount of work that can be obtained from a given quantity of entropy is a maximum when it is delivered from a Carnot engine, but there is a limit to the amount, set by the absolute zero of temperature. On putting T_0 equal to zero in Equation (8.14), it can be seen that the ultimate amount of work, W_u, that can be obtained from a quantity of entropy ΔS, supplied at a temperature T, is given by

$$W_u = - T \Delta S \qquad (8.15)$$

This may be regarded as the absolute work value of the quantity of entropy supplied at a temperature T and embodies F. C. Frank's slogan: temperature is the mechanical equivalent of entropy.

When the state of a given system is changed by the absorption of a quantity of heat, the quantity of heat is actually determined by finding the amount of (adiabatic) work that must be done on the system to produce the same change of state. The properties of a quantity of heat are not, however, completely specified by this value. For example, as Equation (8.14) shows, equal quantities of heat are not, in general, capable of performing the same amount of work when supplied to a Carnot engine in which the reservoir temperatures vary, although the difference in temperature remains constant. In contrast, equal quantities of entropy supplied to the working substance of a Carnot engine always produce equal amounts of work for a fixed difference in the temperature of the two reservoirs, whatever the individual values of the reservoir temperatures.

8.4 The determination of changes in entropy

The entropy change of a system in a given reversible process can be calculated from Equation (8.10) in terms of the heat absorbed by the system and the thermodynamic temperature. In a reversible adiabatic process the entropy change must be zero. However, Equation (8.14) shows that the calculation of the entropy change in a reversible isothermal process can be carried out without measuring quantities of heat: values of the entropy change can be calculated as the work that would be done in a Carnot cycle, using the particular system as the working substance, when the temperatures of the two reservoirs differ by 1 K. This calculation needs only a knowledge of the equation of state.

8.5 The entropy form of the first law of thermodynamics

When a system undergoes any infinitesimal process between initial and final equilibrium states, the change in internal energy, dU, is related to the heat absorbed by the system q and the work done on the system w by

$$dU = q + w \qquad (8.16)$$

When the process is reversible, it was shown in Section 4.6 that w may be written in the form

$$w = \sum_i X_i \, dx_i \tag{8.17}$$

where X_i is a generalised force and x_i the corresponding generalised displacement. Rearranging Equation (8.11) shows that, for a reversible process,

$$q_R = dQ_R = T dS$$

Therefore, combining Equations (8.11), (8.16) and (8.17) gives

$$dU = T dS + \sum_i X_i dx_i \tag{8.18}$$

Equation (8.18) is known as the entropy form of the first law, and, at first sight, it might appear that it applies to reversible processes only. However, all the quantities appearing in Equation (8.18) are, in fact, functions of the state of the system. Therefore, on going from an initial equilibrium state i to a final equilibrium state f, the changes in U, T, S, X_i and x_i depend only on the states i and f and not on the nature of the process by which the change was effected. Equation (8.18) applies, therefore, to any infinitesimal process between equilibrium states. However, only for a reversible process is it true that

$$q = T dS \text{ and } w = \sum_i X_i \, dx_i$$

When the process is irreversible, Equation (8.18) is still valid, but a term-by-term interpretation is not possible: in such a situation $q \neq T dS$ and

$$w \neq \sum_i X_i \, dx_i$$

For example, when a closed hydrostatic system undergoes an infinitesimal reversible process,

$$w = -p dV$$

so that Equation (8.18) becomes

$$dU = T dS - p dV \tag{8.19}$$

Equation (8.19) is valid for any infinitesimal process between two equilibrium states, but only when the process is reversible is it possible to equate $T dS$ with q and $-p dV$ with w.

8.6 Entropy and irreversible processes

It has been shown in Section 8.2 that the entropy S is a state function and so it is natural to ask what features distinguish a reversible process from one that is irreversible, since the change in the entropy of the system is the same for both.

Consider a system of any complexity which is allowed to undergo an irreversible process that takes it from an equilibrium state i to an equilibrium state f and then returns to state i by means of a reversible process.

If the system has only two independent coordinates, X and x, the reversible process can be represented by the line R in the X—x plane (Figure 8.3). Since X and x are not necessarily descriptive of the system during the irreversible process, this process is represented by the hatched line I in Figure 8.3.

The process iIfRi is a cyclic process that is not completely reversible and, therefore, from the inequality of Clausius,

$$\oint \frac{q}{T} < 0 \tag{8.20}$$

where q is the heat absorbed by the system from a source (reservoir) at a temperature T. Equation (8.20) may be written

$$\int_{iIf} \frac{q}{T} + \int_{fRi} \frac{q}{T} < 0 \tag{8.21}$$

where the first integral is evaluated for the irreversible part of the process and the second during the reversible part. For the reversible part of the process,

$$\int_{fRi} \frac{q}{T} = - \int_{iRf} \frac{q}{T}$$

so that relation (8.21) may be written

$$\int_{iIf} \frac{q}{T} < - \int_{fRi} \frac{q}{T} = \int_{iRf} \frac{q}{T}$$

or

$$\int_{iIf} \frac{q}{T} < \int_{iRf} \frac{q}{T}$$

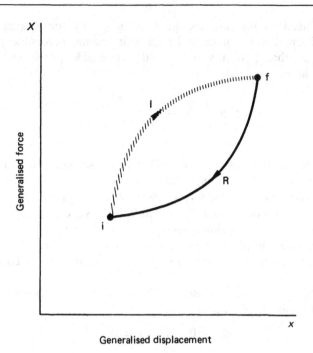

Figure 8.3 *To determine the entropy change in an irreversible process*

Now the integral

$$\int_{iRf} \frac{q}{T}$$

is the difference in the entropy of the system in the equilibrium states i and f — that is,

$$\int_{iRf} \frac{q}{T} = S_f - S_i$$

Therefore,

$$\int_i^f \frac{q_I}{T} < S_f - S_i \qquad (8.22)$$

or, when the process is infinitesimal,

$$\frac{q_I}{T} < dS \qquad (8.22a)$$

Here q_I is used to indicate the heat absorbed by the system in an infinitesimal irreversible process. In an infinitesimal reversible process $dS = q_R/T$, so that both reversible and irreversible processes can be included in the expressions

$$dS \geq \frac{q}{T} \; ; \; S_f - S_i \geq \int_i^f \frac{q}{T} \tag{8.23}$$

where the inequality refers to irreversible processes and the equality to reversible processes. Equations (8.23) indicate one clear distinction between reversible and irreversible processes. When a system goes from an initial equilibrium state i to a final equilibrium state f, the change in entropy of the system has a single value whatever the nature of the process. For all reversible processes this change is given by $\int q/T$, but for all irreversible processes $\int q/T$ is always less than the change in the entropy of the system.

When the system under consideration has an adiabatic boundary, q is zero and relations (8.23) become

$$dS \geq 0 \; ; \; S_f - S_i \geq 0 \tag{8.24}$$

This important result is known as the law of increase of entropy, and states that the entropy of a thermally isolated system can never decrease. When such a system undergoes a reversible process, its entropy remains unchanged, but when it undergoes an irreversible process (which includes all natural processes), its entropy increases. For a system that is isolated from its surroundings relations (8.24) must, of course, hold, but there is the additional condition that the internal energy must remain constant, whether the process is reversible or irreversible.

Relations (8.24) are very useful when a number of systems interact with each other but with no others. These interacting systems can then be treated as parts of an isolated system. If these 'part' systems are labelled A, B, C,..., relation (8.24) becomes

$$\Delta S_A + \Delta S_B + \Delta S_C + \cdots \geq 0$$

When all the processes that take place are reversible, the total entropy of the parts remains constant — that is,

$$\Delta S_A + \Delta S_B + \Delta S_C + \ldots = 0$$

In particular, when only two systems A and B interact, as in Figure 8.4, and all processes are reversible,

$$\Delta S_A + \Delta S_B = 0$$

or

$$\Delta S_A = -\Delta S_B \tag{8.25}$$

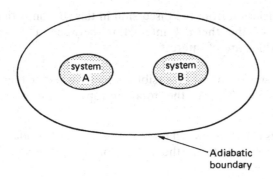

Figure 8.4 *Two interacting systems A and B that do not interact thermally with the rest of the universe*

When irreversibility is present,

$$\Delta S_A + \Delta S_B > 0 \tag{8.26}$$

Relations (8.25) and (8.26) can be used to give a second distinction between reversible and irreversible processes. Consider a system A that interacts with well-defined surroundings B, so that A and B can be treated as isolated from the remainder of the universe, as in Figure 8.5. When the system goes from an equilibrium state i to an equilibrium state f and then returns to equilibrium state i, it suffers no change in entropy, whatever the nature of the processes; the process is, in fact, cyclic, so that there is no change in the value of any of the coordinates. When all the processes are reversible, Equation (8.25) indicates that the total entropy change of the surroundings is also zero. However, when the processes are irreversible, relation (8.26) shows that the change in entropy of the surroundings is positive.

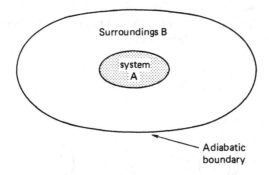

Figure 8.5 *A system A with well-defined surroundings B*

A final conclusion from the discussion in this section is that there are two measures of the thermal interaction between a system and its surroundings which are of interest:

1   One is conserved in any reversible process, whether there is a work interaction or not. This is the total entropy of the system and its surroundings.

2   The other is conserved when there is an adiabatic boundary around the region of interest, whatever the nature of the processes. This is the quantity called heat.

8.7 Maxwell's relations

Thermodynamic coordinates may be classed as primitive or non-primitive. Primitive coordinates are those that are directly measurable, such as pressure and temperature, while non-primitive coordinates, such as entropy and internal energy, must be calculated. In theoretical work the entropy function is important because it can be used when a heat transfer takes place reversibly, rather than the quantity of heat itself, which is not a state function. However, to obtain relationships involving only quantities that are measured experimentally, it is then necessary to relate entropy changes to changes in the values of the primitive coordinates. This can be done by means of equations known as Maxwell's relations, which will now be derived for a closed hydrostatic system. They are readily obtained for other types of system (see, for example, Section 12.2).

In any infinitesimal process between two equilibrium states of a closed hydrostatic system, the change in the internal energy of the system is given by Equation (8.19) — that is, by

$$dU = TdS - pdV$$

This equation shows that it is the extensive variables S and V that govern changes in U; to the first order in dU the changes in the intensive variables can be neglected. From equation (8.19)

$$\left(\frac{\partial U}{\partial S}\right)_V = T \text{ and } \left(\frac{\partial U}{\partial V}\right)_S = -p$$

A second differentiation gives

$$\frac{\partial^2 U}{\partial V \partial S} = \left(\frac{\partial T}{\partial V}\right)_S \text{ and } \frac{\partial^2 U}{\partial S \partial V} = -\left(\frac{\partial p}{\partial S}\right)_V$$

Since U is a real function, the order of differentiation is immaterial and, therefore,

$$\left(\frac{\partial T}{\partial V}\right)_S = -\left(\frac{\partial p}{\partial S}\right)_V \qquad \text{(M4)}$$

This is one of Maxwell's relations [1]. It does not describe a particular process, but gives a relation between the entropy and the primitive coordinates when the system is subject to certain constraints. Further, (M4) is valid only when p, S, V and T are well-defined — that is, for equilibrium states and reversible processes.

The other Maxwell relations may be obtained in a similar manner. If V and T are treated as the independent coordinates, S may be written as $S = S(T,V)$ and then

$$dS = \left(\frac{\partial S}{\partial T}\right)_V dT + \left(\frac{\partial S}{\partial V}\right)_T dV$$

Substituting for dS in Equation (8.19) gives

$$dU = T\left[\left(\frac{\partial S}{\partial T}\right)_V dT + \left(\frac{\partial S}{\partial V}\right)_T dV\right] - p dV$$

and, on rearranging,

$$dU = T\left(\frac{\partial S}{\partial T}\right)_V dT + \left[T\left(\frac{\partial S}{\partial V}\right)_T - p\right] dV$$

Since dU is a complete differential,

$$\left(\frac{\partial}{\partial V}\right)_T \left[T\left(\frac{\partial S}{\partial T}\right)_V\right] = \left(\frac{\partial}{\partial T}\right)_V \left[T\left(\frac{\partial S}{\partial V}\right)_T - p\right]$$

or

$$T\frac{\partial^2 S}{\partial V \partial T} = T\frac{\partial^2 S}{\partial T \partial V} + \left(\frac{\partial S}{\partial V}\right)_T - \left(\frac{\partial p}{\partial T}\right)_V$$

that is,

$$\left(\frac{\partial p}{\partial T}\right)_V = \left(\frac{\partial S}{\partial V}\right)_T \qquad \text{(M3)}$$

Similarly, taking p and S as independent coordinates gives

$$\left(\frac{\partial V}{\partial S}\right)_p = \left(\frac{\partial T}{\partial p}\right)_S \qquad \text{(M2)}$$

while taking p and T as independent coordinates gives

$$\left(\frac{\partial V}{\partial T}\right)_p = \left(\frac{\partial S}{\partial p}\right)_T \qquad \text{(M1)}$$

Equations (M1), (M2), (M3) and (M4) are the four Maxwell relations for a closed hydrostatic system.

■ EXAMPLE

Q. A reversible heat engine operates between reservoirs at temperatures of 400 K, 300 K and 200 K, respectively. During one complete cycle of the engine, the working substance extracts 1200 J of heat from the reservoir at 400 K and rejects 100 J of heat to the reservoir at 200 K. Calculate the heat absorbed per cycle from the reservoir at 300 K and the work done on the working substance per cycle.

A. This problem may be solved by applying the first and second laws of thermodynamics to one cycle of the engine. Let the heat absorbed per cycle be Q_1, Q_2, Q_3 from the reservoirs at temperatures 400 K, 300 K and 200 K, respectively. Then, if W is the work done on the working substance per cycle, the first law is

$$Q_1 + Q_2 + Q_3 + W = 0$$

since, in one cycle, there is no change in the internal energy of the working substance. Substituting the given values, the above equation becomes

$$1200 + Q_2 + (-100) + W = 0 \qquad \text{(8.27)}$$

Since the working substance undergoes a cyclic process, its entropy change is zero, and, since all the processes are reversible, the second law may be written in the form $\Sigma Q/T = 0$, where Q is the heat absorbed per cycle from the reservoir at a temperature T. For the engine considered this gives

$$\frac{Q_1}{400} + \frac{Q_2}{300} + \frac{Q_3}{200} = 0$$

or, substituting the given values,

$$\frac{1200}{400} + \frac{Q_2}{300} + \frac{(-100)}{200} = 0 \qquad (8.28)$$

Solving Equation (8.28) for Q_2 and substituting this value in Equation (8.27) gives

$$Q_2 = -750 \text{ J}$$
$$W = -350 \text{ J}$$

The negative sign for the value of Q_2 indicates that the heat is rejected to the reservoir at 300 K; the negative sign for the value of W indicates that the engine does work on its surroundings.

■ EXERCISES

1   A kettle of water is just simmering on a ring of an electric cooker. To evaporate 0.5 kg of liquid water when the temperature is 373 K and the pressure is 1 standard atmosphere takes 0.32 kW h of electrical energy. Estimate the increase in entropy of 1 kg of water in changing from liquid water to steam under the conditions given.

2   A reversible engine operates between reservoirs at temperatures of 400 K, 300 K and 200 K, respectively. During some integral number of complete cycles the engine extracts 1200 J of heat from the reservoir at 400 K and performs 200 J of work upon the surroundings. Calculate the heat absorbed from each of the other reservoirs.

3   Show the states through which the working substance of a Carnot engine passes during one cycle, using a temperature–entropy graph. When temperature is plotted as the ordinate, what does the area enclosed by this curve represent?

4   An electric current of 1 A flows for 10 s in a resistor of resistance 25 Ω which is submerged in a large volume of water, the temperature of which is 280 K. Calculate

(a) the change in the entropy of the resistor, and
(b) the change in the entropy of the water.

5   The equilibrium states of a closed hydrostatic system may be represented by the Gibbs U–V–S surface, where the symbols have their usual meanings. What quantities are represented by $(\partial U/\partial S)_V$ and $(\partial U/\partial V)_S$ on this surface?

6 An isolated hydrostatic system is contained in a right circular cylinder with its axis horizontal and is divided into two parts, 1 and 2, by a frictionless piston. If T_1, T_2 and p_1, p_2 are the temperatures and pressures, respectively, in the two parts, find the conditions for thermodynamic equilibrium of the system

   (a) when the piston provides a barrier that is impermeable and adiabatic, and
   (b) when it provides a barrier that is impermeable and diathermic.

Note

1 The M stands for Maxwell; the numbers are those given to the equations in Maxwell's book *Theory of Heat*, page 169 (London: Longmans, Green and Company, first edition, 1871).

9
The ideal gas and thermodynamic temperature

The essential theory of classical thermodynamics has now been established, apart from the third law of thermodynamics, which will be considered in Chapter 15. Much of the remainder of this book consists of applications of the theory established so far to the description of particular situations and the characterisation of particular systems. However, from a practical viewpoint, the measurement of thermodynamic temperature was left in Chapter 7 in a rather unsatisfactory state: thermodynamic temperature could only be determined by means of a device whose performance was a close approximation to that of a Carnot engine. This situation will now be remedied by determining the equation of state for ideal gases.

9.1 The equation of state for an ideal gas

The ideal gas is defined from the limiting behaviour of the real gases at vanishingly small pressures. For 1 mol of molecules of a real gas, occupying a volume V_m when its pressure is p,

$$\lim_{p \to 0} pV_m = f(\theta) \tag{9.1}$$

which is Boyle's law. The internal energy U satisfies the equation

$$\lim_{p \to 0} \left(\frac{\partial U}{\partial V} \right)_\theta = 0 \tag{9.2}$$

which is Joule's law. Here θ is an empirical temperature and $f(\theta)$ is a universal function of temperature only. These two laws are based on observations made at constant temperature and need no particular temperature scale for their determination. For ideal gases Equations (9.1) and (9.2) become, respectively,

$$pV_m = f(\theta) \tag{9.3}$$

and

$$\left(\frac{\partial U}{\partial V}\right)_\theta = 0 \tag{9.4}$$

When any closed hydrostatic system undergoes an arbitrary infinitesimal process between two equilibrium states,

$$dU = TdS - pdV \tag{9.5}$$

where T is the thermodynamic temperature and S is the entropy. Therefore,

$$\left(\frac{\partial U}{\partial V}\right)_T = T\left(\frac{\partial S}{\partial V}\right)_T - p \tag{9.6}$$

The entropy may be eliminated from this equation by using the Maxwell relation

$$\left(\frac{\partial p}{\partial T}\right)_V = \left(\frac{\partial S}{\partial V}\right)_T \tag{M3}$$

so that

$$\left(\frac{\partial U}{\partial V}\right)_T = T\left(\frac{\partial p}{\partial T}\right)_V - p \tag{9.7}$$

Equation (9.7) is valid for any closed hydrostatic system; it relates the internal energy to directly measurable quantities and is known as the energy equation.

If the closed hydrostatic system is, in fact, an ideal gas,

$$\left(\frac{\partial U}{\partial V}\right)_\theta = \left(\frac{\partial U}{\partial V}\right)_T = 0$$

since the result holds at constant temperature. Equation (9.7) then becomes

$$\left(\frac{\partial p}{\partial T}\right)_V = \frac{p}{T} \tag{9.8}$$

Integrating Equation (9.8) gives

$$\ln p = \ln T + F(V) \tag{9.9}$$

where $F(V)$ is a function of integration [1]. Equation (9.9) may be written

$$p = T \phi(V) \tag{9.10}$$

where $\phi(V)$ is also an unknown function of V.

Using the thermodynamic temperature function, Boyle's law for an ideal gas may be written in the form

$$pV_m = \chi(T) \text{ or } p = \frac{\chi(T)}{V_m} \tag{9.11}$$

where $\chi(T)$ is an unknown function of temperature which is the same for 1 mol of molecules of all ideal gases.

When temperature is kept constant, it follows from Equation (9.10) that

$$p \propto \phi(V)$$

and from Equation (9.11) that, for 1 mol of molecules,

$$p \propto 1/V_m$$

Therefore,

$$\phi(V) \propto \frac{1}{V_m} \text{ or } \phi(V) = \frac{R}{V_m}$$

and the constant of proportionality R has the same value for 1 mol of molecules of all ideal gases. Substituting for $\phi(V)$ in Equation (9.10) gives as the equation of state for 1 mol of molecules of ideal gas, when temperature is defined by the thermodynamic temperature function,

$$pV_m = RT \tag{9.12}$$

The universal constant R is known as the *molar gas constant* or, more commonly, as the *gas constant*. Its value is obtained from measurements of p and V_m when the temperature is that of the triple point of water; the result is $R = 8.1345 \text{ J K}^{-1} \text{ mol}^{-1}$. Equation (9.12) is known as the ideal gas equation.

If V is the volume occupied by n mol of molecules of ideal gas, $V_m = V/n$ and Equation (9.12) becomes

$$pV = nRT \tag{9.13}$$

Representations of Equation (9.13) under various conditions are given in Figure 9.1.

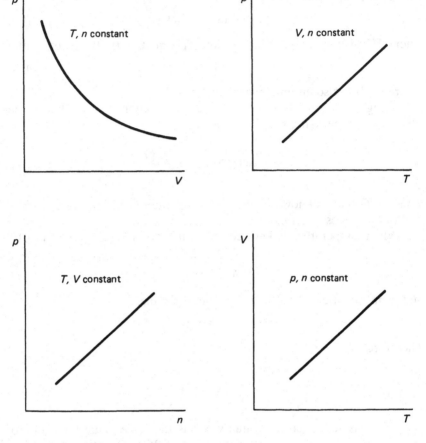

Figure 9.1 *Representations of ideal gas behaviour*

For real gases, the equations corresponding to (9.12) and (9.13) are, respectively,

$$\lim_{p \to 0} pV_m = RT \qquad (9.14)$$

and

$$\lim_{p \to 0} pV = nRT \qquad (9.15)$$

It follows from Equation (9.14) that the coefficient A in the expansion

$$pV_m = A + Bp + Cp^2 + \cdots$$

is given by

$$A = RT$$

When p and n are kept constant, Equations (9.13) and (9.15) become, respectively,

$$\frac{V}{T} = \text{constant, for ideal gases} \qquad (9.16)$$

$$\lim_{p \to 0} \frac{V}{T} = \text{constant, for real gases} \qquad (9.17)$$

Early experiments by Dalton (1802) and, independently and more precisely, by Gay-Lussac (1802) showed that, between the freezing point and normal boiling point of water, the fractional change in volume of all gases, at constant pressure, is the same, a result they referred to as the law of uniformity of gaseous expansion. This result has its explanation in Equation (9.17).

9.2 Mixtures of ideal gases

It will be shown in Chapter 17 that the behaviour of ideal gases is shown by a model gas in which the molecules have negligible size and exert negligible attractive forces on each other. Consequently, in a mixture of ideal, non-reacting gases, the molecules of the different gases move independently of each other and the properties of the mixture will be a combination of those which each gas would have if it alone were present. An ideal gas mixture consisting of non-reacting components satisfies the law of molar conservation — namely,

$$n = \sum_{1}^{r} n_i$$

where the numbers of moles of molecules of the components before mixing are $n_1, n_2, \ldots, n_r$, respectively, and n is the total number of moles of molecules after mixing. For an ideal gas $p = nRT/V$, so that, if the components are separately confined in a volume V, at the same temperature T as in the mixture, they exert so-called partial pressures p_i, given by $p_i = n_i RT/V$. Then, if the law of molar conservation holds,

$$n = \frac{pV}{RT} = \sum n_i = \sum \frac{p_i V}{RT} = \frac{V}{RT} \sum p_i \text{ (constant } T, V)$$

or

$$p = \sum p_i \qquad (9.18)$$

This is Dalton's law of partial pressures.

Alternatively, the ideal gas equation may be written $V = nRT/p$. Then, if the volume V_i of component i is measured at a given temperature and pressure,

$$V_i = n_i \, RT/p$$

and, if the law of molar conservation holds,

$$V = \sum V_i \quad \text{(constant } T,p) \qquad (9.19)$$

This is Leduc's law of partial volumes. Both Dalton's and Leduc's laws are limiting laws which are obeyed by real gases only in the limit of vanishingly small pressures.

9.3 Gas thermometers

The basic requirements of thermometry in terms of thermodynamic temperatures are the specification of a reproducible fixed point, having an arbitrarily chosen numerical value, and the specification of thermodynamically sound methods of relating the numerical values of other systems in equilibrium to that of the defining fixed point.

The first requirement was satisfied in 1954 when the Tenth General Conference of Weights and Measures adopted the temperature of the triple point of water as the defining fixed point and assigned to it the value of 273.16 K (see Section 7.7).

Equations (9.14) and (9.15) can be used to satisfy the second requirement, since they show that thermodynamic temperatures may be determined from measurements of the pressure and volume of real gases, extrapolated to vanishingly small pressures. This is the basis for the gas thermometer, which is the principal instrument for determining thermodynamic temperature.

The Chappuis version of a gas thermometer operating at constant volume is shown schematically in Figure 9.2. In this instrument the gas is contained in the bulb B, which is connected by a capillary tube C to the mercury column, shown shaded. To maintain the gas at constant volume, the height of the reservoir R is varied until the surface of the mercury just touches the tip of the pointer P_1. The enclosed space above the mercury

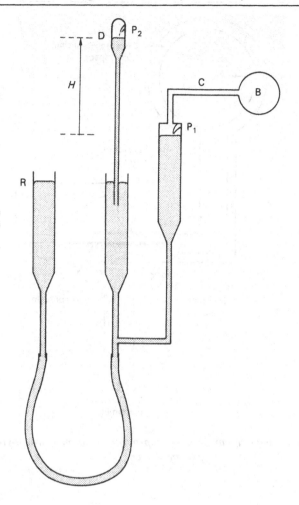

Figure 9.2 *Constant-volume gas thermometer of Chappuis*

surface is known as the 'dead space' or 'nuisance volume', as the gas in it is not, in general, at the temperature of the gas in B. Tube D is a Torricellian barometer. When the height of this tube is adjusted so that the surface of the mercury in it touches the tip of the pointer P_2, the pressure of the gas in B equals the pressure exerted by the mercury column of height H, which is given by the difference in the heights of the pointers P_1 and P_2.

In modern developments the thermometric gas is linked to the gas above the manometer by a stiff diaphragm, as in Figure 9.3. When this diaphragm is made into one plate of a capacitor, the other plate being rigid, movement of the diaphragm can be detected through changes in

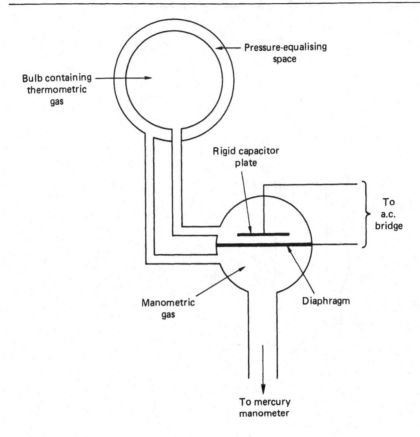

Figure 9.3 *Schematic diagram showing improvements to the constant-volume gas thermometer*

capacitance. To measure the pressure of the thermometric gas, the height of the mercury reservoir is varied until the displacement of the diaphragm is zero. Then the thermometric gas and manometer gas exert the same pressure. If the manometric gas is allowed to surround the bulb of the thermometer, no correction is needed to allow for changes in the volume of the thermometric gas. Further, since the mercury surface no longer defines the volume of the thermometric gas, the accuracy of the pressure measurement can be improved by using wide tubes for the mercury.

The gas thermometer is not usually used in routine measurements, as it is clumsy and slow-acting and the active volume is large. Further, precision gas thermometry is very demanding, as a number of corrections must be applied before reliable thermodynamic temperatures are obtained: corrections are needed for the effect of the dead space, the expansion of the thermometer bulb B and thermomolecular pressure effects [2], but the most important correction is for the non-ideality of the gas. Equations (9.14) and (9.15) suggest that this last correction may be made by repeating the measurement at different gas pressures and extrapolating to zero pressure. However, this turns out to be an inexact procedure, so that, in practice, temperature measurements are made, using gas at a finite pressure, usually 1000 mmHg for 1 mol of molecules at the triple-point temperature, and then a correction is applied to give the value applicable to vanishingly small pressures.

9.4 Gas thermometer correction

One method of obtaining the correction for the non-ideality of the gas in a constant-volume gas thermometer is as follows. Using Equation (9.14), thermodynamic temperatures are obtained from

$$T = T_3 \lim_{p \to 0} \frac{p}{p_3} \quad (V \text{ constant}) \tag{9.20}$$

where p_3 is the pressure of the gas at the temperature T_3, the triple-point temperature of water. The measurement made in practice, however, is of a temperature T_g measured on an empirical constant-volume gas scale of temperature defined by

$$T_g = T_3 \frac{p}{p_3} \quad (V \text{ constant}) \tag{9.21}$$

To develop a correction, assume that the behaviour of the (real) gas is sufficiently well represented by the equation

$$pV_m = A + Bp \tag{9.22}$$

where A and B are functions of temperature only. Then, from Equation (9.21),

$$T_g = T_3 \frac{pV_m}{p_3 V_m} = T_3 \frac{(A + Bp)}{(A_3 + B_3 p_3)} \tag{9.23}$$

where A_3 and B_3 are the values of the coefficients at the temperature T_3. Now $A = RT$ and $A_3 = RT_3$, so that, substituting in Equation (9.23) gives

$$T_g = T_3 \frac{(RT + Bp)}{(RT_3 + B_3 p_3)}$$

or

$$T = T_g \left[1 + \frac{B_3 p_3}{RT_3} \right] - \frac{Bp}{R} \qquad (9.24)$$

Equation (9.24) allows T to be found in terms of T_g.

9.5 The International Temperature Scale

The disadvantages of gas thermometry for routine use have been mentioned in Section 9.3. However, it is important to be able to determine thermodynamic temperatures and, to assist in this, the International Practical Temperature Scale was introduced in 1927. This scale of temperature is easily and accurately reproducible and gives, as nearly as possible, thermodynamic temperatures. The scale is based on the assigned values of the temperature of a number of reproducible equilibrium states (defining fixed points) and on standard interpolation instruments, calibrated at these fixed points, and using agreed formulae to relate the indications of the standard instruments to International Practical Temperature. This International Temperature Scale, as it is now called, has, with the exception of the triple-point temperature of water, assigned values of its fixed points which are the best current estimates of the (true) thermodynamic temperature (in kelvins) at each fixed point.

Table 9.1 gives the fixed points. These are established by realising specified equilibrium states between phases of pure substances — that is, through melting points, boiling points and triple points. The numerical values are determined at national standards laboratories, largely using gas thermometry. With the exception of the triple point of water, values in Table 9.1 are not defined but are experimental values that may change slightly as gas and other fundamental thermometer techniques improve. The values in Table 9.1 are those of the 1990 revision of the scale [3]. This revised scale is known as the *International Temperature Scale 1990*, abbreviated to ITS-90. Temperatures measured on this scale are denoted T_{90}.

Instructions are given in the specification of the scale for the construction and calibration of the interpolation instruments. From 0.65 K to 5 K, T_{90}/K is obtained from measurements of the equilibrium vapour pressure of ^{3}He and ^{4}He, using specified equations. As there are no fixed points between 5 K and 13.8033 K, a constant-volume gas thermometer is specified for the range 3–24.5561 K, calibrated at a temperature between 3 K and 5 K, using a helium vapour pressure, and at the triple points of hydrogen and neon.

Table 9.1 Defining fixed points of the ITS-90. Except for triple points, the assigned values of temperature are for equilibrium states at a pressure of 1 standard atmosphere (101 325 Pa), unless otherwise stated

Equilibrium state	Assigned value of International Temperature, T_{90}/K
Triple point of equilibrium hydrogen[a]	13.8033
Boiling point of equilibrium hydrogen[a] at a pressure of 33 321.3 Pa	17.035
Boiling point of equilibrium hydrogen[a] at a pressure of 101 292 Pa	20.27
Triple point of neon	24.5561
Triple point of oxygen	54.3584
Triple point of argon	83.8058
Triple point of mercury	234.3156
Triple point of water	273.16: exact by definition
Melting point of gallium	302.9146
Freezing point of indium	429.7485
Freezing point of tin	505.078
Freezing point of zinc	692.677
Freezing point of aluminium	933.473
Freezing point of silver	1234.93
Freezing point of gold	1337.33
Freezing point of copper	1357.77

[a]The hydrogen has its equilibrium *ortho–para* composition at the relevant temperature.

In the range of temperatures between 13.8033 K and 1234.93 K temperatures are determined from the ratios between the resistance of a prescribed platinum resistance thermometer at a temperature T_{90} and its resistance at the triple point of water. For temperatures in the range 13.8033–273.16 K calibration is obtained from the triple points of hydrogen, neon, oxygen, argon and mercury and two temperatures near 17 K and 20 K. For the remainder of the temperature range the calibration uses fixed points taken from the melting point of gallium and the freezing points of indium, tin, zinc, aluminium and silver.

At temperatures above the freezing point of silver, ITS-90 is defined by means of the Planck radiation law (see Sections 18.12 and 18.16), taking as a reference the radiation emitted from a black-body cavity at the freezing point of silver, gold or copper according to choice.

9.6 Practical thermometry

For precision temperature measurement the instruments specified in ITS-90 should be used, but in less exacting work a number of other instruments are used. These latter must be calibrated against a standard instrument so as to indicate thermodynamic temperatures as closely as possible. Some commonly used instruments will now be considered.

Resistance thermometers measure changes in temperature through changes in electrical resistance. Platinum is widely used in precision measurements; it can be obtained in a very pure condition and this makes for good reproducibility. In the best current practice a resistance thermometer is constructed as a four-terminal resistor, the wire is mounted and maintained in a strain-free condition, the sensing element is surrounded by a sheath which is filled with a dry gas and hermetically sealed, and the dimensions are made as small as possible. The latter two features help the thermometer quickly to reach thermal equilibrium with the system whose temperature is being measured. To measure the resistance a form of Kelvin bridge or a modifed Wheatstone bridge is usually used.

Another form of resistance thermometer uses a thermistor, which is a thermally sensitive resistor whose temperature coefficient of resistance is negative and, at room temperature, can be numerically much greater than those of common metals. Thermistors are usually made of a mixture of oxides of copper, manganese and nickel. The resistance R is given by

$$R = R_\infty \exp(B/T) \tag{9.25}$$

where R_∞ is the limiting value of R at very high temperatures and B is a constant. Thermistors can be made very small and, therefore, have a rapid response to changes in the temperature being measured.

When two wires made, respectively, of different metals are joined to produce a continuous circuit and the junctions are maintained at different temperatures, an e.m.f. is produced which causes a current to flow. This is the *Seebeck effect* and the e.m.f. is known as a *thermoelectric e.m.f.* The magnitude of the e.m.f. depends on the nature of the metals and on the temperature difference between the junctions. For many pairs of metals the e.m.f. E is well represented by an equation of the form

$$E = a_1(\delta T) + a_2(\delta T)^2 + a_3(\delta T)^3 \tag{9.26}$$

where the as are constants and δT is the temperature difference between the junctions. When the e.m.f. is used to determine temperature differences, the device is called a *thermocouple*. In precision work the Pt–10% Rh/Pt thermocouple is widely used. Pure and strain-free material gives good reproducibility up to about 1700 K. Thermocouples can have small junctions and, therefore, respond quickly to changes in temperature. When high precision is needed, the e.m.f. is measured with a potentiometer. The approximate voltage sensitivity and working range for some commonly used thermocouples are given in Table 9.2.

The expansion of mercury in a glass envelope is still widely used as a means of measuring temperature. Relative to glass, mercury expands fairly uniformly with thermodynamic temperature and, when the envelope is made of of vitreous silica instead of glass, readings are reproducible to a few hundredths of a kelvin, although, for such high accuracy, several corrections must be applied. If mercury is replaced by other liquids, such as ethyl alcohol, the range of use can be changed.

For temperatures in excess of about 1200 K measurements are made using the radiation emitted by a system by virtue of its temperature. (See Section 18.16.)

Table 9.2 Properties of some commonly used thermocouples

Thermocouple	Room temperature thermal e.m.f./mV K^{-1}	Useful temperature range/K
Copper/constantan	0.040	100–600
Iron/constantan	0.050	100–1000
Chromel/alumel	0.040	100–1500
Pt–10% Rh/Pt	0.006	100–1700

Constantan = 60% Cu; 40% Ni.
Chromel = 90% Ni; 10% Cr.
Alumel = 95% Ni, plus Al, Si, Mn.
Values from Kaye and Laby (1973).

■ EXAMPLE

Q. A capillary tube of length L, closed at both ends, contains ideal gas at a pressure p_0 when the temperature is uniform and equal to T_0. If the temperature variation along the tube is changed to

$$T = T_0 \, e^{-cx}$$

where x is the distance measured from the end of the tube at a temperature T_0 and c is a constant, show that the new equilibrium pressure p is given by

$$p = p_0 \frac{cL}{(e^{cL}-1)}$$

A. When the temperature of the gas is uniform, the uniformity of the pressure is achieved by having the gas uniformly distributed throughout the tube. After the temperature variation has been imposed on the gas, the pressure must still be uniform when equilibrium is reached. This is achieved by a redistribution of mass, although, since the gas constitutes a closed system, the total mass of gas must remain constant.

Let the number of moles of gas molecules be n and the area of cross-section of the capillary tube be A. Then, applying the ideal gas equation to the situation where the temperature is uniform and equal to T_0, gives

$$p_0 A L = n R T_0$$

When the temperature gradient is imposed, let an element dx of the tube, at a distance x from the end where the temperature is T_0, contain dn' mol of molecules. Applying the ideal gas equation to this element gives

$$p \, A \, dx = RT \, dn'$$

and substituting for T in terms of x gives $p \, A dx = R \, T_0 \, e^{-cx} dn'$. Now p, A, T_0 and R are constants, so that separating the variables gives

$$\frac{pA}{RT_0} \int_0^L e^{cx} dx = \int_0^n dn'$$

or

$$\frac{pA}{RT_0 c} (e^{cL} - 1) = n$$

Now

$$\frac{nRT_0}{A} = p_0 L$$

so that

$$p = p_0 \frac{cL}{(e^{cL} - 1)}$$

■ EXERCISES

1   Derive the equation of state of 1 mol of molecules of an ideal gas when temperature is measured using the efficiency universal temperature function, defined by Equation (7.18).

2   A sample of an ideal gas is contained in a vessel of constant volume V_0 that is connected to another of constant volume V by a capillary of negligible volume. When both vessels are at a temperature T_0 the equilibrium pressure

in the system is p_0. Show that the equilibrium pressure p when the vessel of volume V has its temperature changed to T is given by

$$p = p_0 \left(\frac{T}{\alpha T + (1-\alpha)T_0} \right)$$

where $\alpha = V_0/(V+V_0)$.

3 Show that for an ideal gas the cubic expansivity is the reciprocal of its thermodynamic temperature.

4 Show that, when n moles of molecules of an ideal gas undergo a reversible isothermal expansion from an equilibrium state with coordinates p_i, V_i to one with coordinates p_f, V_f, the heat Q absorbed by the gas and the work W done on the gas are given by

$$Q = -W = nRT \ln \left(\frac{V_f}{V_i} \right) = nRT \ln \left(\frac{p_i}{p_f} \right)$$

5 Is it possible to decide whether an ideal gas is a chemically simple one, or a mixture of chemically different but non-interacting gases, by an investigation of the equation of state?

6 The bulb of a certain constant volume gas thermometer has a volume V and is connected by a capillary tube to a mercury manometer where there is a dead space of $V/100$ above the mercury. In a particular series of measurements the gas in the dead space remains at 20°C. If p is the measured pressure when the bulb of the thermometer is maintained at 100°C, determine the value of p' that the pressure would have been if all the gas had been at a temperature of 100°C. Assume that the gas is ideal and neglect the expansion of the bulb.

Notes

1 Integration of a partial differential equation introduces an arbitrary function of the variable held constant in the differentiation, rather than a constant of integration.

2 When the bore of the tube C is so small that it is comparable to the mean free path of the gas molecules (see Section 17.5), a difference in pressure can exist between the ends of the tube when equilibrium is achieved.

3 The boiling point of water substance under a pressure of 1 standard atmosphere is now believed to be 99.975°C.

10

Thermodynamic potential functions

A very important result in classical thermodynamics is the law of increase of entropy, discussed in Section 8.6. This law states that, when a change takes place in a system with an adiabatic boundary, the entropy S of the system always increases when the process is irreversible and remains the same when the process is reversible. When, in addition, there is no work interaction, so that the system is isolated, there is the further condition that the internal energy U of the system is unchanged.

The above result is particularly valuable in situations where the behaviour of a system is influenced only by its immediate surroundings. Then the system and such surroundings may be treated as a larger isolated system to which the law of increase of entropy can be applied. For example, the behaviour of a small block of metal immersed in a lake is not influenced by changes taking place beyond the lake, at least in short times: the metal block and a suitable portion of the surrounding water can then be treated as an isolated system.

The law of increase of entropy is the most general statement of the second law of thermodynamics. However, its very generality means that, in practical terms, it is often not the most convenient form of statement in a particular situation. This will be illustrated by considering the application of the law of increase of entropy to the behaviour of a system that interacts with a single heat reservoir, this being a good representation of many real situations.

10.1 The law of increase of entropy for a system interacting with a single reservoir

Let any closed system with well-defined surroundings undergo any process that takes it from an equilibrium state i to an equilibrium state f. If the change is infinitesimal and, during the course of it, the system absorbs a quantity of heat q and has work [1] done on it of amount w, the change in the internal energy U of the system is given by

$$dU = q+w \tag{10.1}$$

From the law of increase of entropy, the associated entropy change is given by

$$dS + dS_0 \geqslant 0 \tag{10.2}$$

where dS is the entropy change of the system and dS_0 is that of the surroundings. The equality sign holds when the process is reversible.

If it is assumed that the surroundings consist of a reservoir at a uniform and constant temperature T_0, all changes taking place in the surroundings as a result of heat transfer can be treated as reversible, so that dS_0 may be written

$$dS_0 = -q/T_0$$

Substituting for dS_0 in Equation (10.2) gives

$$dS - \frac{q}{T_0} \geqslant 0 \tag{10.3}$$

or, substituting for q from Equation (10.1),

$$dU - T_0\, dS \leqslant w \tag{10.4}$$

To obtain information about finite processes, Equation (10.4) must be integrated. Since w is path-dependent, this integration can only be carried out when something is known of the conditions to which the system is subject. A number of important situations will now be considered.

10.2 Adiabatic processes

Let the only constraint on the behaviour of the system be that the process is adiabatic. Then $q = 0$ and, from Equation (10.3),

$$dS \geqslant 0 \tag{10.5}$$

Therefore, as a result of an adiabatic process, the entropy of any system increases when the process is irreversible and remains constant when the process is reversible. When, in addition, $w = 0$, the internal energy is unchanged. These results are simply restatements of the law of increase of entropy and have been discussed in Section 8.6.

10.3 Isothermal processes

Now consider a closed system that has a diathermic boundary. When thermal equilibrium is achieved, the temperature T of the system is uniform and equal to T_0, that of the heat reservoir that forms the surroundings (see Figure 10.1). Equation (10.4) may then be written

$$d(U - TS) \leqslant w; \quad (T = T_0) \tag{10.6}$$

or, putting $U - TS$ equal to F,

$$dF \leqslant w; \quad (T = T_0) \tag{10.7}$$

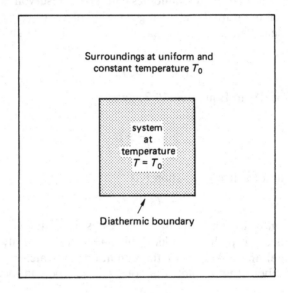

Figure 10.1 *A system in thermal contact with a reservoir at a temperature* T_0

The quantity F is a function of the state of the system having the dimensions of energy and is known as the *Helmholtz function* of the system. Equation (10.6) shows that in an isothermal process, with the temperature of the system equal to that of the surroundings, dF is less than dU by the amount of TdS. This latter term is the energy that is unavailable for the performance of work because of entropy changes in the system: in an isothermal process the internal energy of the system is changed by both the absorption of heat and the performance of work.

For an infinitesimal reversible isothermal process

$$dF = w; \quad (T = T_0)$$

and in a finite reversible isothermal process, in which the system goes from an equilibrium state i to an equilibrium state f,

$$F_f - F_i = W; \quad (T = T_0) \tag{10.8}$$

where W is the total work of all types done on the system during the process. Therefore, in a finite reversible isothermal process the total work done on the system by the surroundings is equal to the increase in the Helmholtz function of the system. Conversely, when the system does work on the surroundings in a reversible isothermal process, the total amount of work done is equal to the decrease in the Helmholtz function of the system [2]. Since F is a state function, the total work done on the system in a reversible isothermal process depends only on the initial and final equilibrium states of the system. Therefore, provided that the process is reversible, it is not essential that it be carried out at constant temperature, but simply that the initial and final temperatures have the same value — namely, that of the surroundings.

When an isothermal process is irreversible,

$$F_f - F_i < W; \quad (T = T_0) \tag{10.9}$$

i.e. in an irreversible isothermal process the increase in the Helmholtz function is less than the total work done on the system. Conversely, when the system does work on the surroundings, the amount of work done is less than if the same change were brought about by a reversible process.

Many chemical reactions take place at constant temperature and with practically no work interaction. In these conditions Equation (10.9) becomes

$$F_f - F_i < 0 \tag{10.10}$$

i.e. the Helmholtz function decreases. The amount of this decrease may be taken as a measure of the work done by the forces causing the reaction. Equation (10.10) also shows that a reaction will occur only if the Helmholtz function of the product is less than the sum of Helmholtz functions of the reactants.

Because the only natural changes that occur in a system in contact with a heat reservoir at the same temperature do so in the direction of decreasing F, the Helmholtz function is known as a thermodynamic potential. Equations (10.8) and (10.9) show that the capability of a system to do work under isothermal conditions is determined by the Helmholtz function of the system rather than its internal energy. In any isolated region of the universe energy is conserved but natural isothermal processes can occur, provided that there is a reduction in the total value of the Helmholtz function for the region. In the mechanics of single particles moving in conservative force fields, energy and the Helmholtz function are identical, but once finite bodies are considered, a distinction is necessary.

The reason that moving objects and coiled springs can do work is because they possess Helmholtz free energy.

When a process is not isothermal, the behaviour of the Helmholtz function is obtained from the differential

$$dF = d(U - TS) = dU - TdS - SdT \qquad (10.11)$$

Substituting for dU from Equation (10.4) gives

$$dF \leqslant w + T_0 dS - TdS - SdT \qquad (10.12)$$

This equation is not, in general, very useful: when there is a change in temperature of the system, the relation between the work interaction and the Helmholtz function becomes complicated.

10.4 **Isothermal, isobaric processes**

Now consider a closed hydrostatic system with a boundary that is both diathermic and free to move without friction — that is, both work and thermal interactions may occur between the system and its surroundings. Further, let the surroundings be not only a heat reservoir at a uniform and constant temperature T_0, but also exert a uniform and constant pressure p_0 (see Figure 10.2). Equilibrium is achieved when the temperature T of the system is uniform and equal to T_0 and, in addition, when the pressure p in

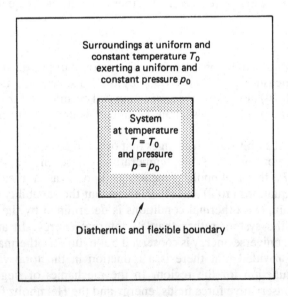

Figure 10.2 *A system in contact with a reservoir that is at a temperature T_0 and exerts a uniform pressure p_0*

the system is uniform and equal to p_0. If it is assumed that the only work interaction between the system and its surroundings is pressure–volume or displacement work and this work is considered to be carried out reversibly, then

$$w = -p_0 dV$$

Substituting for w in Equation (10.4) gives

$$dU - T_0 dS \leqslant -p_0 dV$$

or, since $T = T_0$ and $p = p_0$,

$$dU + pdV - TdS \leqslant 0$$

or

$$d(U + pV - TS) \leqslant 0; \qquad (T = T_0 \quad p = p_0) \qquad (10.13)$$

The function $U + pV - TS$ has the dimensions of energy and is a function of the state of the system, although it is only useful under the conditions $T = T_0$, $p = p_0$ and, in that sense, is a property of both system and surroundings. This function is known as the *Gibbs function* of the system, symbol G. Equation (10.13) may then be written

$$dG \leqslant 0; \qquad (T = T_0 \quad p = p_0) \qquad (10.14)$$

and shows that, when a closed hydrostatic system undergoes a change such that its temperature and pressure have the same values, respectively, as for the surroundings, the value of G decreases when the process is irreversible and remains constant when the process is reversible.

The Gibbs function is also a thermodynamic potential, since, under the conditions specified in Equation (10.14), natural changes in the system that occur do so in the direction of decreasing G.

10.5 Useful work and availability

In almost all previous discussions of closed hydrostatic systems only displacement work has been considered. Since this work involves merely pushing back the surroundings, it is often called useless work, although such work is, in fact, the only type that a closed hydrostatic system can perform without a 'machine' of some kind (see Section 4.9).

In addition to interacting with the surroundings, a system can also be coupled to a second system, which will be called the body, and which will be treated as having an adiabatic boundary, so that the only interaction between the system and the body is a work interaction. The system can then have work done on it by the body as well as by the hydrostatic surroundings. Then, in an infinitesimal process, the work interaction w consists of two parts: w', the non-displacement work interaction with the body, and w'', the useless work interaction with the hydrostatic surroundings: $w = w' + w''$.

Consider a closed hydrostatic system in surroundings that are at a uniform and constant temperature T_0 and exert a uniform and constant pressure p_0 on the system. The heat q absorbed by the system in any infinitesimal process must satisfy Equation (10.3) — namely,

$$q \leqslant T_0 dS$$

so that the change dU in the internal energy of the system is

$$dU = q + w \leqslant T_0 dS + w$$

If the displacement work is carried out reversibly,

$$w'' = -p_0 dV$$

and

$$w' \geqslant (dU + p_0 dV - T_0 dS) \qquad (10.15)$$

Since p_0 and T_0 are constant, Equation (10.15) may be written

$$w' \geqslant d(U + p_0 V - T_0 S)$$

or

$$w' \geqslant dB$$

(10.16)

where $B = U + p_0 V - T_0 S$ and is known as the availability of the system. Clearly, B is, in fact, a property of the system and its surroundings, since it involves p_0 and T_0, but it is a useful function only when p_0 and T_0 are constant.

w' is the non-displacement work done in a given change in given surroundings; it has a maximum value for a reversible process. When this work is done by the system on the body, this is indicated by dB being negative — that is, the availability decreases — and the work is useful work.

To illustrate this result, the useful work will be determined for a simple system for which it can be calculated directly—namely, an electric cell in which gas production occurs. If the cell communicates with the atmosphere by means of a weightless, frictionless piston, as in Figure 10.3, the pressure throughout the cell will be constant at p_0, the pressure of the surroundings. The cell may be kept at a constant temperature T_0 by being placed in a constant-temperature bath. The non-displacement work w' may be carried out reversibly by placing a potentiometer in the external circuit so that only an infinitesimal current flows. There is then a negligible potential drop across the internal resistance of the cell and the p.d. of the cell approaches closely its maximum value.

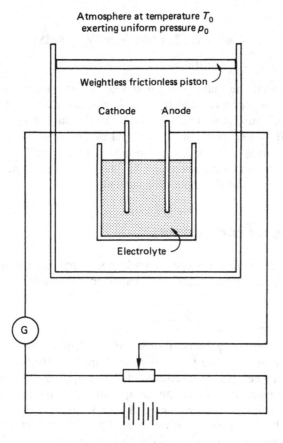

Figure 10.3 *An example illustrating the performance of both useful and useless work*

Let an infinitesimal charge dZ flow from the anode to the cathode in the external circuit and let the volume of gas evolved in the process be dV. In this process non-displacement work is done on the cell of amount EdZ, where E is the e.m.f. of the cell. The displacement work done on the cell is given by $-p_0 dV$, so that the total work done on the cell is given by

$$w = EdZ - p_0 dV$$

If the cell is in thermal equilibrium with the surroundings, its temperature is T_0, that of the surroundings. The first law for the infinitesimal process may then be written

$$dU = T_0 dS + w$$
$$= T_0 dS + EdZ - p_0 dV \tag{10.17}$$

so that

$$w' = EdZ = (dU + p_0dV - T_0dS)$$

$$= dB \tag{10.18}$$

This confirms that the maximum amount of non-displacement work that may be done in a process in which the surroundings are characterised by a constant temperature and pressure is equal to the change in the availability. When dB is negative, the non-displacement work is done on the surroundings and is useful work. Note that in this example, where $p=p_0$ and $T = T_0$, the availability is equal to the Gibbs function. This is not generally true (see the Example on page 141).

10.6 **Natural coordinates**

When a closed hydrostatic system undergoes any infinitesimal process between two equilibrium states, the change in the internal energy dU is given by

$$dU = TdS - pdV$$

provided that the only work interaction is displacement work. This equation shows that, to the first order, changes in U are determined only by changes in the extensive coordinates entropy, S, and volume, V; changes in the intensive coordinates temperature, T, and pressure, p, may be ignored. Because of this property, S and V are known as the natural coordinates of U. The partial derivatives of U with respect to the natural coordinates give what are known as the conjugate coordinates. For a closed hydrostatic system the conjugate coordinates are

$$T = \left(\frac{\partial U}{\partial S}\right)_V ; \; -p = \left(\frac{\partial U}{\partial V}\right)_S \tag{10.19}$$

The natural coordinates for the Helmholtz function F and the Gibbs function G can be found from expressions for dF and dG, respectively.

The Helmholtz function is given by

$$F = U - TS$$

so that

$$dF = dU - TdS - SdT$$

But, from the first law, when the only work interaction is displacement work,

$$dU = TdS - pdV$$

giving

$$dF = -pdV - SdT \qquad (10.20)$$

Consequently, the natural coordinates for F are T and V. When a closed hydrostatic system undergoes a process at constant V and T, there is no change in the Helmholtz function.

The Gibbs function is given by

$$G = U + pV - TS$$

so that

$$dG = dU + pdV + Vdp - TdS - SdT$$

Substituting for dU from the first law gives

$$dG = Vdp - SdT \qquad (10.21)$$

from which it follows that the natural coordinates for G are p and T. Clearly, when a closed hydrostatic system undergoes a process at constant temperature and pressure, there is no change in G. This result is useful when considering phase changes and many chemical reactions.

■ EXAMPLE

Q. Derive an expression for the useful work done when 1 mol of molecules of ideal gas expands reversibly at constant temperature T_0 from a pressure p_i to a pressure p_f in surroundings that exert a uniform and constant pressure p_0 and have a uniform and constant temperature T_0.

A. To keep the gas in equilibrium when its pressure differs from that of its surroundings and no frictional forces are present, the system must be coupled to a body; it is this coupling that allows useful work to be performed.

Equation (10.16) shows that the useful work done by the sytem is equal to the decrease in the availability B of the system when the surroundings exert a uniform and constant pressure p_0 and have a uniform and constant temperature T_0. The work w' done by the body on the system is given by

$$w' = dB = dU + p_0dV - T_0dS$$

For an ideal gas the internal energy U may be written $U = U(T)$, so that, when T is constant, dU is zero and then

$$w' = p_0dV - T_0dS; \quad (T = T_0)$$

Now the entropy S may be treated as a function of the temperature and pressure of the system:

$$S = S(T,p)$$

so that

$$dS = \left(\frac{\partial S}{\partial T}\right)_p dT + \left(\frac{\partial S}{\partial p}\right)_T dp$$

In an isothermal process dT is zero and, using Maxwell relation (M1) (page 114), the above equation becomes

$$dS = \left(\frac{\partial S}{\partial p}\right)_T dp = -\left(\frac{\partial V}{\partial T}\right)_p dp; \; (T = T_0)$$

The equation of state for an ideal gas is

$$pV_m = RT$$

where V_m is the molar volume and R is the gas constant. Therefore,

$$\left(\frac{\partial V}{\partial T}\right)_p = \frac{R}{p}$$

so that

$$w' = T_0 \frac{R}{p} dp + p_0 dV$$

In a finite process, when the volume of the gas changes from V_{mi} to V_{mf}, W' is given by

$$W' = R T_0 \int_{p_i}^{p_f} \frac{dp}{p} + p_0 \int_{V_{mi}}^{V_{mf}} dV$$

$$= R T_0 \ln\left(\frac{p_f}{p_i}\right) + p_0(V_{mf} - V_{mi})$$

which, using the equation of state, may be written

$$W' = RT_0 \ln\left(\frac{p_f}{p_i}\right) + p_0 R T_0 \left(\frac{1}{p_f} - \frac{1}{p_i}\right) \tag{10.22}$$

When useful work is done on the body by the system, W' is negative.

For example, let $p_0 = 10^5$ Pa, $T_0 = 300$ K, $p_i = 2 \times 10^5$ Pa and $p_f = 1.5$ Pa, so that the system is expected to perform useful work on the body. Substitution of these values in Equation (10.22) gives

$$W' = 8.315 \times 300 \times \ln \frac{1.5}{2} + 10^5 \times 8.135 \times 300 \times \left(\frac{1}{1.5 \times 10^5} - \frac{1}{2 \times 10^5}\right)$$

taking R as 8.315 J K^{-1} mol^{-1}. Evaluating this expression gives $W' = -301.9$ J. As expected, W' is negative, indicating that the system does useful work on the body.

The problem may also be approached directly by considering the total work done by the gas in the expansion and subtracting the useless work done against the surroundings. In an infinitesimal process at the constant temperature T_0, the total amount of work w done on the system is given by

$$w = dF = d(U - T_0 S) = dU - T_0 dS$$

Now the entropy form of the first law is

$$dU = T_0 dS - pdV$$

so that

$$w = -pdV$$

where p is the pressure of the gas. This work w is composed of the displacement work w'' done by the surroundings and the work w' done by the body: $w = w' + w''$. The work w'' done by the surroundings is given by

$$w'' = -p_0 dV$$

Therefore, w' is given by

$$w' = -pdV + p_0 dV \qquad (10.23)$$

Substituting for pdV from the entropy form of the first law gives

$$w' = dU + p_0 dV - T_0 dS$$

the right-hand side of which is dB, the change in the availability of the system. Substituting for p from the equation of state into Equation (10.23) gives

$$w' = -RT_0 \frac{dV}{V} + p_0 dV$$

which can then be integrated to give

$$W' = -RT_0 \ln\left(\frac{V_{mf}}{V_{mi}}\right) + p_0(V_{mf} - V_{mi})$$

$$= RT_0 \ln \left(\frac{p_f}{p_i}\right) + p_0 RT_0 \left(\frac{1}{p_f} - \frac{1}{p_i}\right)$$

as before.

■ EXERCISES

1   Show that, for a closed hydrostatic system where displacement work is the only work interaction, the internal energy U, Helmholtz function F, Gibbs function G and temperature T are related through the following equations:

$$U = F - T\left(\frac{\partial F}{\partial T}\right)_V$$

$$G = F - V\left(\frac{\partial F}{\partial V}\right)_T$$

which are both known as the Gibbs–Helmholtz equation.

2   The Massieu function J for the system of Exercise 1 is defined by the equation

$$J = S - \frac{U}{T}$$

where S is the entropy. Show that

$$dJ = \frac{U}{T^2} dT + \frac{p}{T} dV$$

3   A certain polymer sample may be treated as perfectly elastic and has the following approximate equation of state when it is in the form of a cylinder of length L, under a tension F at a temperature T:

$$F = (aT - bT^2)(L - L_0)$$

L_0 is the unstretched length and a and b are constants. Derive expressions for the change in internal energy and in the Gibbs function of the specimen when it is extended from its unstretched length L_0 to $2L_0$, reversibly and isothermally.

Notes

1   It should be emphasised that w is the total amount of work done on the system. If, for example, the system is an electric cell that generates gas, work may be done by the cell in pushing back the atmosphere in addition to driving an electric current around a circuit.

2   The function F has a relation to the total work W done on the system similar to that which the internal energy has to the sum $Q + W$, where Q is the heat absorbed by the system. For this reason, F is sometimes known as the free energy for the isothermal process or, simply, as the free energy.

11

Heat capacity

Doing work on a system and the absorption of heat by a system are simply different ways of changing the internal energy of a system. For a pure work interaction the work done on a system is specified solely in terms of the lowering or raising of a weight with respect to a reference level. However, when the process concerned is reversible, the work done may also be obtained from the generalised force and displacement coordinates of the system.

The heat absorbed by a system in a pure thermal interaction may be determined by finding the amount of adiabatic work needed to bring about the same changes in the system. For a system that is not undergoing a phase change, the response to a transfer of energy in the form of heat or work, or both, is a rise in temperature. For a given system this temperature rise depends on the quantity of energy transferred, the nature of the system and the nature of the process during which the transfer takes place. The study of this energy response leads to the introduction of another response function of the system, known as the heat capacity.

11.1 Definition of heat capacity

Let any closed hydrostatic system absorb a quantity of heat Q without undergoing a phase change. Experiment shows that, for closed hydrostatic systems of a given mass, subject to given external constraints but composed of different chemical species, the absorption of a fixed quantity of energy produces a temperature rise that is characteristic of the material of which the system is composed [1].

If the absorption of a quantity of heat Q produces a change from an initial equilibrium state with temperature T_i to a final equilibrium state with temperature T_f in a system of constant mass, then $\overline{C}$, defined by

$$\overline{C} = \frac{Q}{T_f - T_i} = \frac{Q}{\Delta T} \tag{11.1}$$

145

is called the *mean heat capacity* of the system over the temperature range T_i to T_f, and its unit is JK^{-1}. The *heat capacity* C at a temperature T is the limiting value of $\overline{C}$ as ΔT tends to zero:

$$C = \lim_{\Delta T \to 0} \frac{Q}{\Delta T} \qquad (11.2)$$

The heat capacity per unit mass is called the *specific heat capacity c* of the system. For a system of constant mass m the specific heat capacity is given by

$$c = \frac{C}{m} = \frac{1}{m} \lim_{\Delta T \to 0} \frac{Q}{\Delta T} \qquad (11.3)$$

and has the unit $J\ kg^{-1}\ K^{-1}$. When a system consists of n moles of a specified entity, a *molar heat capacity* C_m may be defined by

$$C_m = \frac{C}{n} = \frac{1}{n} \lim_{\Delta T \to 0} \frac{Q}{\Delta T} \qquad (11.4)$$

For a particular system the value of the heat capacity depends on the nature of the process in which the heat absorption occurs, since the system may also be able to do work W on its surroundings: from the first law

$$Q = \Delta U - W \qquad (11.5)$$

and, in general, W will be non-zero. For heat capacity to be properly defined, the constraints operative during heat absorption must be specified. If the process is reversible, the path is specified and W becomes a well-defined quantity. A given value of Q then produces a definite change in the internal energy U which is reflected in a definite change in temperature.

The heat capacities corresponding to constancy of the coordinates found in the expression for W are known as the principal heat capacities. For example, a closed hydrostatic system has primitive coordinates pressure, p, volume, V, and temperature, T. Provided that displacement work is the only work interaction, $dW = -pdV$ in a reversible process, so that the principal heat capacities are the heat capacity at constant pressure C_p and the heat capacity at constant volume C_V. These heat capacities are defined, respectively, by

$$C_p = \lim_{\Delta T \to 0} \frac{Q_p}{\Delta T} = \frac{dQ_p}{dT} \qquad (11.6)$$

and

$$C_V = \lim_{\Delta T \to 0} \frac{Q_V}{\Delta T} = \frac{dQ_V}{dT} \tag{11.7}$$

where dQ_p and dQ_V are the heats absorbed by the system in infinitesimal reversible processes at constant pressure and volume, respectively.

The importance of these two heat capacities is that theoretical calculations of solid-phase properties of materials are invariably made for a system at constant volume. Experimentally, however, systems are usually studied at constant pressure, it being very difficult to achieve constancy of volume even for very small changes of temperature (see Exercise 1). It is, therefore, important to be able to relate measurements made at constant pressure to the corresponding values at constant volume, particularly for heat capacities, which are the most easily measured thermal properties that characterise a particular system.

It is found experimentally that, in general, the heat capacity of a system under given constraints is a function of the temperature of the system, and can be treated as constant over small ranges of temperature only.

The discussion in the remainder of this chapter will be restricted to closed hydrostatic systems in which displacement work is the only work interaction, and to processes that are reversible. Closed hydrostatic systems are typical of systems with two degrees of freedom, and the results obtained are readily extended to other such systems.

11.2 Heat capacities of a closed hydrostatic system

For any infinitesimal change of a closed system the first law of thermodynamics is

$$q = dU - w$$

where q is the heat absorbed by the system and w is the work done on the system. For a closed hydrostatic system undergoing a reversible process, where the only work done is displacement work, $w = -pdV$, where p is the pressure exerted by the system and V is its volume. q then behaves as a differential dQ, so that the first law may be written

$$dQ = dU + pdV \tag{11.8}$$

From Equation (11.8) it follows that, under conditions of reversible heat transfer, the principal heat capacities of the system are given by

$$C_V = \frac{dQ_V}{dT} = \left(\frac{\partial U}{\partial T}\right)_V \tag{11.9}$$

and

$$C_p = \frac{dQ_p}{dT} = \left(\frac{\partial U}{\partial T}\right)_p + p\left(\frac{\partial V}{\partial T}\right)_p \qquad (11.10)$$

In a reversible process the change in entropy S is given by $dS = dQ/T$, so that equivalent expressions for the principal heat capacities are

$$C_V = T\left(\frac{\partial S}{\partial T}\right)_V \qquad (11.11)$$

and

$$C_p = T\left(\frac{\partial S}{\partial T}\right)_p \qquad (11.12)$$

For a heat transfer to be closely reversible, the temperature difference between a system and its surroundings must be very small. An estimate of how small this temperature difference must be can be made in the following way. When a quantity of heat Q flows from surroundings at a temperature T to a system at a temperature $T + \delta T$, the change in entropy of the surroundings is $-(Q/T)$, while that of the system is $Q/(T + \delta T)$. The change in entropy of the system and surroundings is, therefore,

$$\frac{Q}{T+\delta T} - \frac{Q}{T}$$

and, in a perfectly reversible process, this should be zero. The total entropy change in the process considered is

$$Q\left(\frac{T-(T+\delta T)}{T(T+\delta T)}\right)$$

and, when T is small, this may be written $-[Q(\delta T/T^2)]$ or $Q\delta(1/T)$. Therefore, for the real process to approach reversibility, $\delta T/T$ must be made smaller than the experimental accuracy.

Equation (11.10) may be written more concisely using the function H, defined by

$$H = U + pV \qquad (11.13)$$

H is a function of state variables, has the dimensions of energy and is known as the *enthalpy* of the system [2].

Because H is a state function, in any infinitesimal process between equilibrium states

$$dH = dU + pdV + Vdp \qquad (11.14)$$

Combining Equations (11.8) and (11.14) shows that in an infinitesimal reversible process

$$dH = dQ + Vdp$$

so that [3]

$$C_p = \frac{dQ_p}{dT} = \left(\frac{\partial H}{\partial T}\right)_p \tag{11.15}$$

11.3 Relations between the principal heat capacities

The first important relations to be examined are the dependence of C_V on V and of C_p on p. Consider C_V first.

In a reversible process

$$C_V = T\left(\frac{\partial S}{\partial T}\right)_V$$

so that

$$\left(\frac{\partial C_V}{\partial V}\right)_T = T\frac{\partial}{\partial V}\bigg)_T \left(\frac{\partial S}{\partial T}\right)_V \tag{11.16}$$

Since S is a state function, the order of differentiation is immaterial and, therefore,

$$\left(\frac{\partial C_V}{\partial V}\right)_T = T\frac{\partial}{\partial T}\bigg)_V \left(\frac{\partial S}{\partial V}\right)_T$$

The entropy may be eliminated from this equation using Maxwell relation (M3) (page 113), giving

$$\left(\frac{\partial C_V}{\partial V}\right)_T = T\frac{\partial}{\partial T}\bigg)_V \left(\frac{\partial p}{\partial T}\right)_V = T\left(\frac{\partial^2 p}{\partial T^2}\right)_V \tag{11.17}$$

The dependence of C_p on p can be examined in a similar manner, the result being

$$\left(\frac{\partial C_p}{\partial p}\right)_T = -T\left(\frac{\partial^2 V}{\partial T^2}\right)_p \tag{11.18}$$

The quantities $(\partial^2 p/\partial T^2)_V$ and $(\partial^2 V/\partial T^2)_p$ may be obtained from the equation of state of the system. For example, if the system consists of n moles of molecules of an ideal gas, the equation of state is

$$pV = nRT$$

where R is the gas constant. Both second partial derivatives are then zero, so that, for an ideal gas,

$$\left(\frac{\partial C_V}{\partial V}\right)_T = \left(\frac{\partial C_p}{\partial p}\right)_T = 0$$

Therefore, for an ideal gas, C_V does not depend on V and C_p does not depend on p. Note, however, that nothing is said about the temperature dependence of either C_V or C_p.

An expression for the difference between the principal heat capacities is readily obtained using Equations (11.11) and (11.12), together with the fact that the entropy is a function of the state of the system and may be expressed in terms of the independent coordinates. A closed hydrostatic system has two independent coordinates, and it is convenient to choose T and V. Then S may be expressed in the form

$$S = S(T, V)$$

so that, in an infinitesimal change,

$$dS = \left(\frac{\partial S}{\partial T}\right)_V dT + \left(\frac{\partial S}{\partial V}\right)_T dV$$

Differentiating with respect to T, keeping p constant, gives

$$\left(\frac{\partial S}{\partial T}\right)_p = \left(\frac{\partial S}{\partial T}\right)_V + \left(\frac{\partial S}{\partial V}\right)_T \left(\frac{\partial V}{\partial T}\right)_p$$

Now, from Equations (11.11) and (11.12),

$$\left(\frac{\partial S}{\partial T}\right)_p = \frac{C_p}{T} \quad \text{and} \quad \left(\frac{\partial S}{\partial T}\right)_V = \frac{C_V}{T}$$

so that

$$C_p = C_V + T\left(\frac{\partial S}{\partial V}\right)_T \left(\frac{\partial V}{\partial T}\right)_p \tag{11.19}$$

It is convenient to eliminate the entropy from Equation (11.19), so that $C_p - C_V$ may be expressed in terms of the primitive coordinates. This may be done using Maxwell relation (M3) (page 113), giving

$$C_p - C_V = T\left(\frac{\partial p}{\partial T}\right)_V \left(\frac{\partial V}{\partial T}\right)_p \tag{11.20}$$

Equation (11.20) enables $C_p - C_V$ to be calculated for a closed hydrostatic system when the equation of state is known. For n moles of molecules of ideal gas

$$\left(\frac{\partial p}{\partial T}\right)_V = \frac{nR}{V} \quad \text{and} \quad \left(\frac{\partial V}{\partial T}\right)_p = \frac{nR}{p}$$

so that

$$C_p - C_V = nR \tag{11.21}$$

and, when n is equal to unity,

$$C_{p,\mathrm{m}} - C_{V,\mathrm{m}} = R \tag{11.22}$$

An alternative expression for $C_p - C_V$ in terms of derivatives may be obtained using the reciprocity relation involving p, T and V — namely,

$$\left(\frac{\partial p}{\partial T}\right)_V \left(\frac{\partial T}{\partial V}\right)_p \left(\frac{\partial V}{\partial p}\right)_T = -1$$

which may be derived as indicated in the Appendix (page 341). Substituting for $(\partial p/\partial T)_V$ in Equation (11.20) gives

$$C_p - C_V = -T\left(\frac{\partial V}{\partial T}\right)_p^2 \left(\frac{\partial p}{\partial V}\right)_T \tag{11.23}$$

From an experimental standpoint this equation is superior to Equation (11.20), as it does not involve a term at constant volume. Equation (11.23) can also be expressed in terms of the cubic expansivity β and the isothermal compressibility κ_T of the system. The defining equations are

$$\beta = \frac{1}{V}\left(\frac{\partial V}{\partial T}\right)_p \quad \text{and} \quad \kappa_T = -\frac{1}{V}\left(\frac{\partial V}{\partial p}\right)_T$$

and substituting in Equation (11.23) gives

$$C_p - C_V = \frac{T V \beta^2}{\kappa_T} \tag{11.24}$$

Three interesting conclusions may be drawn from Equation (11.23).

1 The first two terms on the right-hand side of the equation are necessarily positive and, as $((\partial p/\partial V)_T$ is negative for all known closed hydrostatic systems, $C_p - C_V$ can never be negative — that is, C_p can never be less than C_V.

2   As T tends to zero, $C_p - C_V$ tends to zero, so that, as absolute zero is approached, the values of C_p and C_V should become more nearly the same.
3 C_p and C_V are also equal for any states of the system for which $(\partial V/\partial T)_p$ is zero. An example of such a state is that of water at its maximum density, which occurs at a temperature of about 277 K when the pressure is atmospheric.

   With the help of Equations (11.17), (11.18) and (11.20) (or 11.23) a considerable saving of effort may be effected in the determination of the principal heat capacities of a closed hydrostatic system. If C_p is known for all equilibrium states of the system, Equation (11.20) can be used to calculate C_V for those states, provided that the equation of state is known. There is no need to attempt to measure C_V experimentally; the effort is more usefully expended in determining the equation of state. Further, if C_p is known for one equilibrium state at a temperature T, Equation (11.18) can be used, in conjunction with the equation of state, to calculate C_p for all equilibrium states at the same temperature. The experimental programme is then clear: C_p and C_V can be determined for all equilibrium states of a closed hydrostatic system, provided that the equation of state is known and C_p has been measured for any one state on each isotherm.
   The ratio between the principal heat capacities, C_p/C_V, is also of interest. Using Equations (11.11) and (11.12), the ratio C_p/C_V, denoted by γ, is given by

$$\gamma = \frac{(\partial S/\partial T)_p}{(\partial S/\partial T)_V} \tag{11.25}$$

The entropy can be eliminated, to give an expression in terms of the primitive coordinates by making use of the reciprocity relations

$$\left(\frac{\partial S}{\partial T}\right)_p \left(\frac{\partial T}{\partial p}\right)_S \left(\frac{\partial p}{\partial S}\right)_T = -1 \ ; \ \left(\frac{\partial S}{\partial T}\right)_V \left(\frac{\partial T}{\partial V}\right)_S \left(\frac{\partial V}{\partial S}\right)_T = -1$$

Substituting for $(\partial S/\partial T)_p$ and $(\partial S/\partial T)_V$ into Equation (11.25) gives

$$\gamma = \frac{(\partial S/\partial p)_T \ (\partial p/\partial T)_S}{(\partial S/\partial V)_T \ (\partial V/\partial T)_S} = \frac{(\partial V/\partial p)_T}{(\partial V/\partial p)_S}$$

$(\partial V/\partial p)_T$ is, of course, equal to $-\kappa_T V$, where κ_T is the isothermal compressibility. Defining an *isentropic compressibility* κ_S by the equation

$$\kappa_S = -\frac{1}{V}\left(\frac{\partial V}{\partial p}\right)_S$$

allows γ to be written in the form

$$\gamma = \frac{\kappa_T}{\kappa_S} \tag{11.26}$$

This is a rather unexpected result, and its appearance illustrates the way that thermodynamics is often able to reveal relationships between quantities that, at first sight, appear to be totally unrelated.

11.4 The determination of ΔU and ΔS

Having decided what measurements must be made to determine C_p and C_V for all equilibrium states of a closed hydrostatic system, it is of interest to carry out a similar analysis for the state functions internal energy and entropy.

For a closed hydrostatic system the internal energy U may be treated as a function of the temperature T and volume V. Then, in any infinitesimal process,

$$dU = \left(\frac{\partial U}{\partial T}\right)_V dT + \left(\frac{\partial U}{\partial V}\right)_T dV \tag{11.27}$$

The partial derivatives in Equation (11.27) may be interpreted from a consideration of the first law. Now, in general,

$$dU = TdS - pdV \tag{11.28}$$

so that differentiating with respect to V with T held constant gives

$$\left(\frac{\partial U}{\partial V}\right)_T = T\left(\frac{\partial S}{\partial V}\right)_T - p \tag{11.29}$$

while differentiating with respect to T with V held constant gives

$$\left(\frac{\partial U}{\partial T}\right)_V = T\left(\frac{\partial S}{\partial T}\right)_V$$

or, from Equation (11.11),

$$\left(\frac{\partial U}{\partial T}\right)_V = C_V \tag{11.30}$$

Entropy may be eliminated from Equation (11.29), using Maxwell relation (M3) (page 113), to give

$$\left(\frac{\partial U}{\partial V}\right)_T = T\left(\frac{\partial p}{\partial T}\right)_V - p \tag{11.31}$$

a relation known as the energy equation, and already encountered as Equation (9.7).

It should be noted that Equations (11.30) and (11.31) apply to reversible processes only, even though Equation (11.28) is generally valid.

Combining Equations (11.27), (11.30) and (11.31) gives

$$dU = C_V\,dT + \left[T\left(\frac{\partial p}{\partial T}\right)_V - p\right]dV \tag{11.32}$$

Because internal energy is a state function, the change in U of a system that undergoes a finite change from an equilibrium state with coordinates T_i and V_i to one with coordinates T_f and V_f depends only on the initial and final equilibrium states and not on the process. Therefore, for the purposes of calculation, the real process may be replaced by a notional reversible process that is performed in two stages. First, the temperature of the system is changed from T_i to T_f while the volume of the system is maintained constant with the value V_i. Then the volume of the system is changed from V_i to V_f while the temperature is maintained constant at T_f. The change in internal energy of the system in going from state i to state f is, therefore,

$$U_f - U_i = \int_{T_i,V_i}^{T_f,V_i} C_V\,dT + \int_{V_i,T_f}^{V_f,T_f} \left[T\left(\frac{\partial p}{\partial T}\right)_V - p\right]dV \tag{11.33}$$

The first integral requires a knowledge of C_V as a function of T when the volume is V_i, while the second can be calculated from the equation of state alone.

For a system consisting of n moles of molecules of ideal gas

$$\left(\frac{\partial p}{\partial T}\right)_V = \frac{nR}{V}$$

so that

$$\left(\frac{\partial U}{\partial V}\right)_T = T\left(\frac{\partial p}{\partial T}\right)_V - p = 0$$

i.e. U is a function of temperature only, a result which reflects the contribution of Joule's law to the equation of state. Therefore, for an ideal gas

$$U_f - U_i = \int_{T_i}^{T_f} C_V \, dT \tag{11.34}$$

or $dU = C_V \, dT$, regardless of the type of process that occurs.

The entropy of a closed hydrostatic system may also be treated as a function of temperature and volume. Then, in any infinitesimal process,

$$dS = \left(\frac{\partial S}{\partial T}\right)_V dT + \left(\frac{\partial S}{\partial V}\right)_T dV \tag{11.35}$$

If the process is reversible, $(\partial S/\partial T)_V$ is equal to C_V/T and Maxwell relation (M3) can be used to eliminate the entropy. These two substitutions give

$$dS = \frac{C_V}{T} dT + \left(\frac{\partial p}{\partial T}\right)_V dV \tag{11.36}$$

Since entropy is a state function, the change in entropy of the system when it goes from an equilibrium state i with coordinates T_i, V_i to a state f with coordinates T_f, V_f depends only on the states i and f. Therefore, to calculate the change in entropy, any reversible process linking states i and f can be used. It is convenient to choose a process that consists of an isovolumic part, in which V remains constant at V_i and T changes from T_i to T_f, followed by an isothermal part, in which $T = T_f$ while V changes from V_i to V_f. The change in entropy in the complete process is then

$$S_f - S_i = \int_{T_i,V_i}^{T_f,V_i} \frac{C_V}{T} dT + \int_{V_i,T_f}^{V_f,T_f} \left(\frac{\partial p}{\partial T}\right)_V dV \tag{11.37}$$

The first integral can be evaluated if C_V is known as a function of T when $V = V_i$, and the second can be evaluated if the equation of state is known. For a reversible isothermal process the first integral is zero and

$$(S_f - S_i)_T = \int_{V_i}^{V_f} \left(\frac{\partial p}{\partial T}\right)_V dV \tag{11.38}$$

so that only the equation of state needs to be known.

For a system consisting of n moles of molecules of an ideal gas $(\partial p/\partial T)_V$ is equal to nR/V and then

$$S_f - S_i = \int_{T_i,V_i}^{T_f,V_i} \frac{C_V}{T} dT + \int_{V_i,T_f}^{V_f,T_f} \frac{nR}{V} dV \tag{11.39}$$

Now C_V may be written as $nC_{V,\mathrm{m}}$ and, when the temperature range $T_\mathrm{f} - T_\mathrm{i}$ is such that $C_{V,\mathrm{m}}$ may be treated as a constant, integration of Equation (11.39) gives

$$S_\mathrm{f} - S_\mathrm{i} = nC_{V,\mathrm{m}} \ln\left(\frac{T_\mathrm{f}}{T_\mathrm{i}}\right) + nR \ln\left(\frac{V_\mathrm{f}}{V_\mathrm{i}}\right) \qquad (11.40)$$

11.5 Principles of calorimetry

The essential steps in the determination of the heat capacity of a system are to supply a quantity of heat Q to the system under the required constraints and to measure the resulting change in temperature, ΔT. Then, if $\overline{C}$ is the appropriate mean heat capacity of the system over the temperature range ΔT,

$$\overline{C} = Q/\Delta T$$

In practice, the system may lose heat to the surroundings while the temperature change is occurring. If this loss is of magnitude Q',

$$Q = \overline{C}\,\Delta T + Q' \qquad (11.41)$$

When the heat absorbed by the system is infinitesimal,

$$\mathrm{d}Q = C\mathrm{d}T + \mathrm{d}Q' \qquad (11.42)$$

where C is the heat capacity of the system, under the appropriate constraints, at the temperature T.

Since heat is not measured directly but by the amount of adiabatic work that produces the same change of state, heat capacities are determined experimentally by doing work on the system. The most convenient form of work is that done in moving electrical charge, and so a resistor is included in the system. When a current I is passed through this resistor under a potential difference E for a time t, the work done on the system is EIt, so that

$$IEt = \overline{C}\,\Delta T + Q'$$

or, as t becomes infinitesimal,

$$IE\mathrm{d}t = C\mathrm{d}T + \mathrm{d}Q' \qquad (11.43)$$

which may be written

$$IE = C\frac{\mathrm{d}T}{\mathrm{d}t} + \frac{\mathrm{d}Q'}{\mathrm{d}t} \qquad (11.44)$$

In this equation C includes the response of the resistor. The instrument used for 'measuring heat exchanges' is called a *calorimeter,* and consists essentially of the sample, in a container, if necessary, a resistor and a thermometer, all maintained in controlled surroundings, as shown schematically in Figure 11.1.

To obtain accurate values of C for a given system, it is necessary to eliminate dQ'/dt from Equation (11.44) or to make a satisfactory estimate of it. When the heat loss occurs through a surrounding fluid such as air, the heat loss is largely by convection, although the loss through radiation can be important when the surface has a high absorptivity (see Section 18.10). Convective heat transfer is a complex process, although there are some simple empirical results [4] , and in most experimental situations the value of dQ'/dt is obtained empirically, rather than theoretically.

Adiabatic jacket calorimetry seeks to eliminate dQ'/dt by adjusting the temperature of the surroundings T_s so that it is always equal to the temperature T of the system. Since the rate of heat loss is a function of $T-T_s$, this makes dQ'/dt equal to zero and Equation (11.44) becomes

$$IE = C\frac{dT}{dt} \tag{11.45}$$

An alternative approach is to make dQ'/dt as small as possible by careful calorimeter design and then to measure it. To simplify this procedure, T_s is often kept constant during the measurement, a technique known as *isothermal jacket calorimetry*.

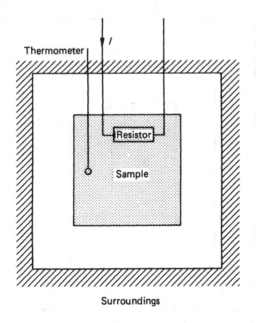

Surroundings

Figure 11.1 *The essential features of a calorimeter*

The essential features of some calorimetric techniques will now be described.

11.5.1 The vacuum calorimeter of Simon and Lange

This is an example of adiabatic jacket calorimetry and is a modification of a method due to Nernst. The technique can be used for solids, liquids and gases.

The specimen Sp (Figure 11.2) is suspended by adiabatic threads inside an opaque, evacuated chamber. Electrical work is done on the specimen, using an insulated resistor R wound on the specimen. Temperature is usually measured with a resistance thermometer whose winding Tm must also be incorporated. Fine wire is used for all electrical connections. Surrounding the specimen is an independently controlled thermostat Th. One junction of a thermocouple (not shown) touches the outer surface of Sp and the other touches the inner surface of Th. The thermostat has a resistor attached to it and the current through this resistor is varied so as to

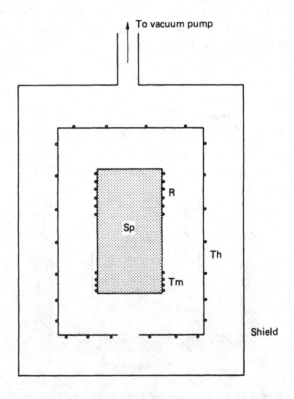

Figure 11.2 *The essential features of the vacuum calorimeter*

keep the thermocouple reading zero. Both Sp and Th are mounted inside an enclosure known as the shield, which is held at constant temperature and which is evacuated. These conditions help to make the control of $T-T_s$ easier by minimising the energy exchanges between Th, Sp and the surroundings and by making the exchange between Th and the surroundings more definite.

In this arrangement dQ' is zero and C is obtained from Equation (11.45); the value of C includes contributions from both the resistor and the thermometer. For solids and liquids this method measures C_p at $p = 0$, but for gases, which need a closed container, it measures C_V (only approximately, because of the small expansion of the container) [5]. The major difficulty in using this method for gases is that, if the container is strong enough to maintain the gas at approximately constant volume as the temperature is increased, the heat capacity of the container is generally much greater than that of the gas, so that the measurement of C_V depends on the measurement of the small difference between the heat capacities of the container when empty and when full.

11.5.2 The cooling-curve method

This is an example of isothermal jacket calorimetry. The method is well suited for solids, when it measures C_p, and for a liquid in equilibrium with its vapour. When used with liquids, it is sometimes called Ferguson's method. The experiment is carried out in two parts.

In the first part the system, which consists of the sample, the incorporated resistor and a thermometer, has electrical work done on it at a constant rate, while it loses energy to surroundings that are kept at a constant temperature T_s. When conditions are steady, dT/dt is zero and Equation (11.44) shows that the rate at which heat is lost to the surroundings is then equal to the rate at which work is done on the system — that is,

$$IE = \frac{dQ'}{dt} \tag{11.46}$$

where I is the current flowing in the resistor under a potential difference E. The steady temperature T_e is recorded and the measurements repeated for other values of the power input $P (= IE)$. A graph of T_e against P has the form shown in Figure 11.3(a).

When the highest temperature of interest has been reached, the current is switched off and the temperature T of the system is measured as a function of time t, giving Figure 11.3(b). For this second part of the experiment Equation (11.44) becomes

$$\frac{dQ'}{dt} = - C_p \frac{dT}{dt} \tag{11.47}$$

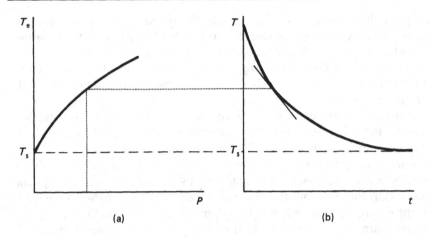

Figure 11.3 *Curves used in the determination of heat capacity by the
cooling-curve method*

Now, for a given sample in constant surroundings, the rate of heat
exchange with the surroundings depends only on $T - T_s$ (see the Appendix to this chapter) and, therefore, must be the same in the two parts of the
experiment when the system has a particular temperature. Taking the
value of P and of dT/dt corresponding to a particular value of the
temperature of the system (shown by the dotted lines in Figure 11.3) gives

$$IE = -C_p \frac{dT}{dt} \qquad (11.48)$$

from which the value of C_p at the chosen temperature may be obtained.

An alternative approach to isothermal jacket calorimetry uses the
arrangement of Figure 11.2, but without the thermostat Th and the
differential thermocouple. With no current flowing in the resistor R, the
temperature T of the sample is measured as a function of time t. The
interval in which these measurements are made is known as the *fore-drift
period*. A current I is then passed under a potential difference E through
the resistor for a time Δt and the temperature is again measured as a
function of time. This is the *after-drift period*. The complete curve of
temperature against time is shown by the full line in Figure 11.4; T_s is the
temperature of the enclosure. The fore- and after-drift curves are then
extrapolated into the middle of the period when the current was flowing
and the temperature difference ΔT at that time is taken as the temperature
rise that would have been produced if there had been no heat exchange
with the surroundings. There is usually a small temperature 'overshoot'

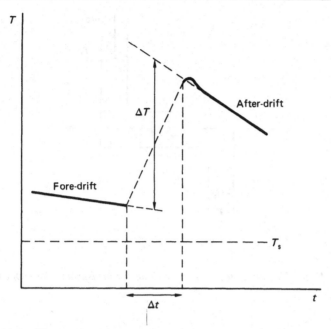

Figure 11.4 *The graph of temperature* T *against time* t *used in the measurement of heat capacity*

which must be allowed for. This method gives a mean heat capacity $\overline{C}_p$ over the temperature range ΔT through the equation

$$IE\Delta t = \overline{C}_p \Delta T \tag{11.49}$$

11.5.3 Continuous-flow calorimetry

A rather different isothermal jacket technique that can be used for fluids is the so-called *method of continuous flow*. The fluid is made to flow through the calorimeter (represented schematically in Figure 11.5) with a constant mass flow rate dm/dt. When conditions are steady, the rate at which energy is supplied to the calorimeter and its contents is equal to the rate at which energy is carried away by the fluid plus the rate at which it is lost to the surroundings. Then, if a current I flows under a potential difference E through the resistor R,

$$IE = \frac{dm}{dt}\overline{c}(T_f - T_i) + \frac{dQ'}{dt} \tag{11.50}$$

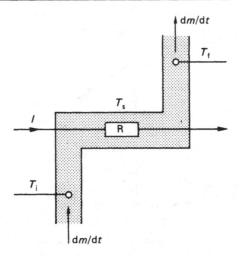

Figure 11.5 *The continuous-flow technique for measuring heat capacity*

Here T_i is the temperature of the fluid at the inlet and T_f its temperature at the outlet; $\bar{c}$ is the mean specific heat capacity over the temperature range T_i to T_f. The temperature rise does not depend on the heat capacity of the calorimeter when conditions are steady, although a correction for heat losses is still necessary. Since the fluid is flowing, there must be a pressure gradient in the calorimeter, so that the heat capacity measured is that at constant pressure, but averaged over the pressure range in the apparatus. As Equation (11.18) shows, this effect is zero for ideal gases and will, therefore, be negligible for dilute real gases.

This method was first applied to gases by Swann (1909) and later, with an improved technique, by Scheel and Heuse (1912), whose apparatus is shown schematically in Figure 11.6. A modern version of the method, as used by Pitzer, is shown in Figure 11.7. Gas flows at a constant rate dm/dt through a U-tube fitted with a thermometer to measure the inlet temperature T_i and two thermometers giving outlet temperatures T_{f1} and T_{f2}. The gas flows past a resistor R through which a current I is flowing under a potential difference E, and then through a gauze pad G to ensure thorough mixing of the gas before the outlet temperatures are measured. A value for the apparent mean specific heat capacity at constant pressure, neglecting heat losses to the surroundings, is calculated, using T_i and T_{f1} and T_i and T_{f2}. These values are

$$\bar{c}_{p1} = \frac{EI}{dm/dt(T_{f1}-T_i)} \tag{11.51}$$

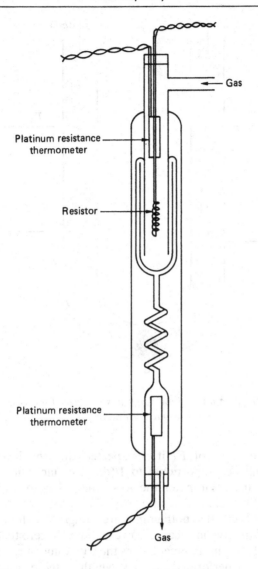

Figure 11.6 *The continuous-flow apparatus of Scheel and Heuse for measuring the heat capacity of a gas. This is a very slightly modified version of Figure 6(b) on page 61 of* An Introduction to Chemical Thermodynamics, *by E. F. Caldin, Oxford University Press, Oxford, 1958*

and

$$\bar{c}_{p2} = \frac{EI}{dm/dt(T_{f2} - T_i)}$$

(11.52)

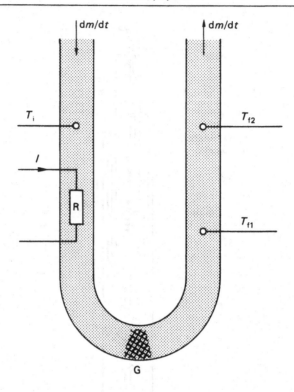

Figure 11.7 *The continuous-flow apparatus of Pitzer for measuring the heat capacity of a gas*

For a given value of EI it is expected that the heat loss to the surroundings will be proportional to $1/(dm/dt)$, since the faster the gas moves through the calorimeter the less time it has to lose heat to the surroundings. When $\bar{c}_{p1}$ and $\bar{c}_{p2}$ are plotted as a function of $1/(dm/dt)$ for constant EI, it is found that both graphs are straight lines for modest ranges of flow rate and that the intercepts, corresponding to zero time of flow, are equal (Figure 11.8). This intercept gives the true value of $\bar{c}_p$. A run with EI equal to zero is performed to see whether there is any spurious Joule–Thomson effect (see Section 16.2).

11.6 Results of heat capacity measurements

Most experimental measurements of heat capacity give the value of C_p for the system studied, but C_V is usually of greater theoretical importance and must be calculated from C_p. This can be done using Equation (11.24). However, when sufficient data are not available for a direct application of

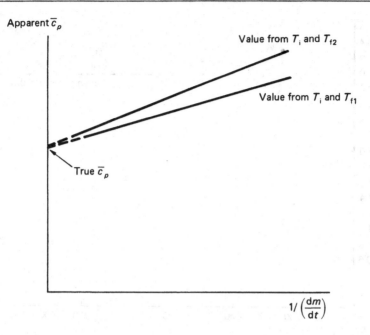

Figure 11.8 *The determination of the true value of the heat capacity of a gas using a continuous-flow technique*

Equation (11.24), a useful procedure is to write the equation in the form

$$C_p - C_V = \alpha C_p^2 T \tag{11.53}$$

where $\alpha = V\beta^2/\kappa_T C_p^2$, since it is found experimentally that, for a large number of solids, α is a constant characteristic of the substance and independent of temperature. Equation (11.53) is known as the Nernst–Lindemann equation. It is also found experimentally that, for different materials, α is proportional to $1/T_m$, where T_m is the normal melting point of the solid. Then Equation (11.53) becomes

$$C_p - C_V = \alpha_0 C_p^2 T/T_m \tag{11.54}$$

where α_0 is a universal constant.

From measurements on a range of pure substances the following generalisations may usefully be made about heat capacities.

1 All heat capacities tend to zero as the temperature approaches absolute zero.

2 For most solids that crystallise in simple systems, such as cubic and hexagonal systems, C_p increases with temperature without ever reaching a constant value. (See Figure 11.9 for the values for 1 mol of atoms of copper.)

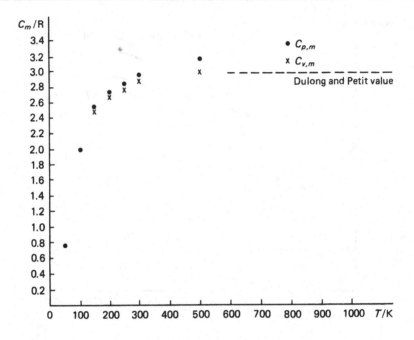

Figure 11.9   *The molar heat capacity* C_m *for 1 mol of atoms of copper*

3   The value of C_V for substances crystallising in simple systems shows a pronounced temperature dependence at low temperatures, where it is almost equal to C_p. As the temperature increases, the value of C_p increases faster than that of C_V and at sufficiently high temperatures C_V tends to a constant value (see Figure 11.9). For a solid system consisting of 1 mol of atoms [6] the high-temperature value of $C_{V,m}$ is approximately $3\,R$, where R is the gas constant, a result known as the Dulong and Petit law.

The term 'high' is a relative one. As Figure 11.10 shows, 200 K is high for lead and 600 K is high for copper, but 700 K is not high for diamond.
4   For the simple substances discussed under (2) and (3) the respective graphs of $C_{V,m}$ against T can be made to coincide by adjusting the scale on the T-axis. This implies that $C_{V,m}$ may be expressed in the form

$$C_{V,m} = F\left(\frac{T}{\Theta}\right) \tag{11.55}$$

where F is the same function for 1 mol of particles of all simple solids and Θ is a constant for a given substance, known as its *Debye temperature* or *characteristic temperature*. Equation (11.55) is really only a good approximation, although it is adequate for calculations over the whole temperature range when only moderate accuracy is required. Detailed examination

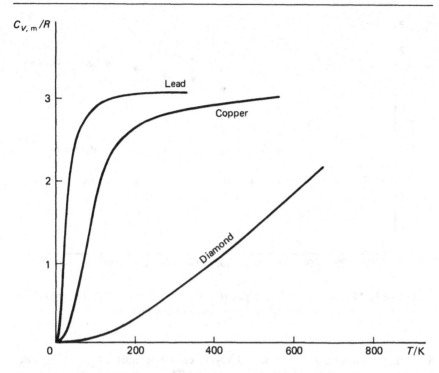

Figure 11.10 *The molar heat capacity* $C_{V,m}$ *(in units of the gas constant R) for lead, copper and diamond, as a function of temperature*

shows that, at very low temperatures, insulators behave rather differently from metals, there being no unique value of Θ.

5 For dilute gases C_p is independent of p and C_V is independent of V, both at constant temperature.

6 For the noble gases and for most metallic vapours $C_{p,m}$ is constant over a wide range of temperature, provided that the gas pressure is low, and has a value of approximately 5 $R/2$ (see Figure 11.11). Under the same condition of low density, $C_{V,m}$ is also constant over a wide range of temperature and is approximately 3 $R/2$. Consequently, $C_{p,m}/C_{V,m}$ is constant and approximately 1.67.

7 Dilute samples of the so-called permanent diatomic gases, such as hydrogen and oxygen, have $C_{p,m}$ approximately equal to 7 $R/2$ at about room temperature, rising slowly with increasing temperature, as in Figure 11.11. $C_{v,m}$ shows a similar temperature dependence, but with a room temperature value of about 5 $R/2$. The ratio $C_{p,m}/C_{V,m}$ is constant around room temperature, being about 1.40, and then decreases as T increases.

At very low temperatures (≈ 50 K) the behaviour of hydrogen (the only diatomic gas that has not condensed by then) changes to that characteristic of monatomic gases.

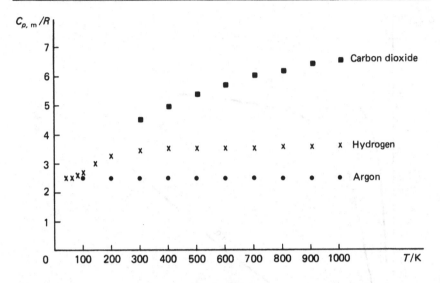

Figure 11.11 *The molar heat capacity at constant pressure $C_{p,m}$ for argon, hydrogen and carbon dioxide at low pressure*

8 For polyatomic gases such as CO_2, even when dilute, $C_{p,m}$, $C_{V,m}$ and $C_{p,m}/C_{V,m}$ vary strongly with temperature. The variation is different for each gas, that for dilute CO_2 being shown in Figure 11.11.

■ EXAMPLES

Q1. At low temperatures the molar heat capacity at constant volume, $C_{V,m}$, of many simple insulating solids is given as a function of temperature T by

$$C_{V,m} = D\left(\frac{T}{\Theta}\right)^3$$

where D has the value 1.94×10^3 J mol^{-1} K^{-1} and Θ, the Debye temperature, is characteristic of the material. For diamond Θ is 1860 K. If 1.00 J of energy is supplied to 1 mol of carbon atoms in the form of diamond (1 mol of diamond), initially at 10.0 K, calculate the final temperature, assuming that the change takes place reversibly at constant volume.

A1. When a closed hydrostatic system, such as a sample of diamond, absorbs a quantity of heat q in an infinitesimal reversible process at constant volume, the temperature rise dT is related to q by

$$q = C_V dT$$

For a system consisting of 1 mol of atoms, the heat Q_V absorbed when the temperature changes from T_i to T_f in a reversible process at constant volume is

$$Q_V = \int_{T_i}^{T_f} C_{V,m} dT$$

Now, for diamond,

$$C_{V,m} = D\left(\frac{T}{\Theta}\right)^3$$

so that

$$Q_V = \int_{T_i}^{T_f} D\left(\frac{T}{\Theta}\right)^3 dT = \frac{D}{\Theta^3}\left[\frac{T_f^4 - T_i^4}{4}\right]$$

Substituting the values for 1 mol of diamond gives

$$1.00 = \frac{1.94 \times 10^3}{1860^3} \frac{[T_f^4 - 10.0^4]}{4}$$

or $T_f = 60$ K.

Q2. Calculate the maximum amount of useful work that may be obtained from 10^6 kg of water at 373 K when a reservoir at 300 K is available. Neglect the change in volume of the water and assume that the specific heat capacity of water at constant volume may be treated as constant over this range and equal to 4200 J kg^{-1} K^{-1}.

A2. When a closed hydrostatic system is in surroundings that exert a constant pressure p_0 and are at a constant temperature T_0, the maximum amount of useful work that may be done by the system in any process is given by the decrease in the availability B, where

$$B = U + p_0 V - T_0 S$$

Here U is the internal energy of the system, V is its volume and S is its entropy. In an infinitesimal process the change in B is

$$dB = dU + p_0 dV - T_0 dS$$

If the process takes place at constant volume, dV is zero, $dS = C_V dT/T$, and $dU = C_V dT$. Then

$$dB = C_V dT - T_0 \frac{C_V dT}{T}$$

and, in a finite process, in which the temperature of the system changes from T_i to T_f,

$$\Delta B = C_V(T_f - T_i) - T_0 C_V \ln\left(\frac{T_f}{T_i}\right)$$

Substituting the values given,

$$\Delta B = 10^6 \times 4200 \,(300 - 373) - 300 \times 10^6 \times 4200 \ln(300/373)$$

$$= -3.2 \times 10^{10} \text{J}$$

ΔB in this process is negative, so that B has decreased, showing that the work is done on the surroundings by the system.

Notes

1   Historically, these energy transfers were made in the form of heat, using water as a standard substance. However, both heat and work produce the same effect in a system and, in current practice, electric work is usually used to investigate the response of systems.

2   Because of its role in determining whether or not processes occur under certain conditions (see Exercise 10), enthalpy is also a thermodynamic potential.

3   Equations (11.10) and (11.15) should be used with care. While Equations (11.6) and (11.12) are generally valid, Equations (11.10) and (11.15) are only valid for systems and surroundings where the only work interaction is displacement work. When other work interactions are present, the heat absorbed at constant pressure is not equal to the change in enthalpy of the system.

4   These are outlined in the appendix to this chapter.

5   In practice, the heat capacity that is measured for liquids is closer to that of the liquid in equilibrium with its vapour, the whole being maintained at approximately constant volume.

6   This means one-half mole of formula units for NaCl and one-third mole of formula units for Fe_2S, and so on.

Appendix: Convection

In many calorimetric arrangements the major heat loss occurs by the motion of the surrounding fluid (usually air) carrying the energy with it.

This process is known as *convection* and the flow of the fluid is called a *convection current*. When these currents are induced by changes in the density of the fluid arising from changes in temperature, the process is known as *natural convection*. This is in contrast to *forced convection,* in which the fluid flow is produced by an external agency.

The convection process is complex, involving first the conduction of heat from the surface of the calorimeter to the fluid and, particularly to the stagnant layer of fluid in contact with the calorimeter surface, followed by heat transfer arising from the motion of the fluid itself, which may be streamline or turbulent.

The combined effect of conduction and convection is incorporated in a *heat transfer coefficient h*, defined by the equation

$$\frac{dQ'}{dt} = - hA(T - T_s) \tag{11.56}$$

where A is the surface area of the calorimeter which is at a temperature T and T_s is the temperature of the main body of the moving fluid. dQ'/dt is the rate of heat transfer. In any particular situation the value of h will depend on whether the surface of the calorimeter is flat or curved, on the relative proportions of the surface that are vertical and horizontal, on the nature of the fluid and on whether the flow is streamline or turbulent.

For forced convection cooling in air of normal laboratory calorimeters, h may usually be taken as a constant. Then Equation (11.56) may be written

$$\frac{dQ'}{dt} = - h_0(T - T_s) \tag{11.57}$$

where h_0 is a constant. Equation (11.57) is usually known as Newton's law of cooling.

When the calorimeter is in still air, so that natural convection controls the heat loss, the following approximate results hold: $h \propto (T - T_s)^{0.25}$ when the air flow is streamline; $h \propto (T - T_s)^{0.33}$ when the air flow is fully turbulent. Substituting these values in Equation (11.56) gives

$$\frac{dQ'}{dt} = h_0' A(T - T_s)^{5/4} \tag{11.58}$$

when the air flow is streamline, a result usually attributed to Dulong and Petit. When the air flow is fully turbulent,

$$\frac{dQ'}{dt} = - h_0'' A(T - T_s)^{4/3} \tag{11.59}$$

■ EXERCISES

1 For copper at room temperature the cubic expansivity is 50×10^{-6} K^{-1} and the isothermal compressibility is 14×10^{-12} Pa^{-1}. Show that, for a temperature rise of 1 K, the external pressure must rise by about 3.6×10^6 Pa to hold a block of copper at constant volume.

2 Two closed hydrostatic systems having constant heat capacities C_1 and C_2 are initially at temperatures T_1 and T_2, respectively. The systems are allowed to reach thermal equilibrium in two different ways:

(a) by heat conduction (irreversible heat transfer);

(b) by operating a Carnot engine between them and extracting work (reversible heat transfer).

Calculate the final equilibrium temperature in the two situations.

3 Close to absolute zero the molar heat capacity at constant pressure $C_{p,m}$ of metallic elements that crystallise in simple systems can be written

$$C_{p,m} = aT^3 + bT$$

where a and b are constants. For 1 mol of atoms of copper at atmospheric pressure $a = 1.16 \times 10^{-5}$ J mol^{-1} K^{-4} and $b = 7.14 \times 10^{-4}$ J mol^{-1} K^{-2}.

Calculate the entropy change of 1 mol of atoms of copper when its temperature changes from 1 K to 9 K at atmospheric pressure.

4 It takes 3 min for the temperature of a certain closed hydrostatic system to fall from 373 K to 333 K when its surroundings are at 283 K. If Newton's law of cooling is obeyed, how long will it take for the temperature to fall from 333 K to 293 K ?

5 A carbon block has a mass of 0.15 kg. When power W is dissipated in it, in given surroundings, it reaches an equilibrium temperature T_e given in the following table:

T_e/K	298	397	442	471	493	514	531
W/W	0	10	20	30	40	50	60

When the block is allowed to cool in the same surroundings with no energy input, the dependence of temperature T on time t is given below:

t/min	0	2	4	6	8	10
T/K	557	504	473	446	434	418

Determine the specific heat capacity at constant pressure of carbon at 460 K.

6 Derive an expression for the variation of the molar internal energy with pressure, at constant temperature, for a gas with equation of state

$$pV_m = RT + Bp$$

where R is the gas constant and B is a function of temperature only.

7 An evacuated vessel with rigid adiabatic walls is connected through a valve to the atmosphere, where the pressure is p_0 and the temperature is T_0. When the valve is opened slightly, air flows into the vessel until the pressure inside it is p_0. Show that, if air is treated as an ideal gas with constant heat capacities, the final temperature of the air in the vessel is γT_0, where γ is C_p/C_V.

8 Show that, for a closed hydrostatic system that has a volume V and exerts a pressure p when its temperature is T, the change in enthalpy H in an infinitesimal process is given by

$$dH = C_p dT + \left[V - T\left(\frac{\partial V}{\partial T}\right)_p \right] dp$$

where C_p is the heat capacity at constant pressure.

9 Derive the following equations for a closed hydrostatic system, where the symbols have their usual meaning:

(a) $C_V = - T\left(\dfrac{\partial^2 F}{\partial T^2}\right)_V$

(b) $C_p = - T\left(\dfrac{\partial^2 G}{\partial T^2}\right)_p$

(c) $C_V = - T\left(\dfrac{\partial p}{\partial T}\right)_V \left(\dfrac{\partial V}{\partial T}\right)_s$

(d) $C_p = T\left(\dfrac{\partial V}{\partial T}\right)_p \left(\dfrac{\partial p}{\partial T}\right)_s$

(e) $C_p - C_V = \left(\dfrac{\partial V}{\partial T}\right)_p \left[p + \left(\dfrac{\partial U}{\partial V}\right)_T \right]$

(f) $TdS = C_V dT + T\left(\dfrac{\partial p}{\partial T}\right)_V dV$

(g) $TdS = C_p dT - T\left(\dfrac{\partial V}{\partial T}\right)_p dp$

(h) $TdS = C_V\left(\dfrac{\partial T}{\partial p}\right)_V dp + C_p\left(\dfrac{\partial T}{\partial V}\right)_p dV$

Equations (f), (g) and (h) are the so-called TdS equations.

10 Show that a closed hydrostatic system, interacting with surroundings at a uniform and constant temperature T_0 that exert a uniform and constant pressure p_0, will only undergo an adiabatic change at constant pressure if this results in a decrease in the enthalpy of the system.

12

The application of thermodynamics to some simple systems

The principles developed in the earlier chapters will now be applied to the systems described in Chapter 5. This will not only produce results of importance, but also illustrate the variety of approaches that can be used to obtain information about thermodynamic systems.

12.1 Closed hydrostatic systems

A closed hydrostatic system is a system of constant mass which exerts a uniform hydrostatic pressure on its surroundings. The primitive coordinates of such a system are pressure p, volume V and temperature T. It is possible to give a complete description of the behaviour of a closed hydrostatic system when it undergoes a reversible process, using equations already obtained. For example, in Section 5.1 it was shown that in an infinitesimal reversible change in volume

$$dV = \beta V dT - \kappa_T V dp$$

where β is the cubic expansivity and κ_T is the isothermal compressibility. However, it is often convenient to derive equations that apply to particular situations and processes. In the next subsection explicit equations will be derived that apply to a closed hydrostatic system that is undergoing a reversible adiabatic, or isentropic, process.

12.1.1 The adiabatic equations

The important quantities describing an isentropic process in a closed hydrostatic system are $(\partial T/\partial p)_S$ and $(\partial T/\partial V)_S$, where S is the entropy of the system. A thermodynamic analysis of an isentropic process should produce expressions for these derivatives in terms of measurable quantities.

175

One starting point for any thermodynamic analysis is the entropy form of the first law. For a closed hydrostatic system undergoing any infinitesimal process between two equilibrium states, this law is

$$dU = TdS - pdV \qquad (12.1)$$

where U is the internal energy of the system. When the process is isentropic, dS is zero and

$$p = - \left(\frac{\partial U}{\partial V} \right)_S \qquad (12.2)$$

Equation (12.2) is useful when considering the behaviour of an ideal gas, for which $dU = C_V dT$, but is not very helpful for other closed hydrostatic systems.

An alternative approach is to start with the appropriate reciprocity relations and try to transform the partial derivatives involved into forms that are related to measurable quantities. In an isentropic process the quantities of interest indicate that the appropriate reciprocity relations are those involving, respectively, T, p and S and T, V and S. The relations are

$$\left(\frac{\partial T}{\partial p} \right)_S \left(\frac{\partial p}{\partial S} \right)_T \left(\frac{\partial S}{\partial T} \right)_p = -1 \qquad (12.3a)$$

and

$$\left(\frac{\partial T}{\partial V} \right)_S \left(\frac{\partial V}{\partial S} \right)_T \left(\frac{\partial S}{\partial T} \right)_V = -1 \qquad (12.3b)$$

Now

$$T \left(\frac{\partial S}{\partial T} \right)_p = C_p$$

so that Equation (12.3a) may be written

$$\left(\frac{\partial T}{\partial p} \right)_S = - \frac{T}{C_p} \left(\frac{\partial S}{\partial p} \right)_T$$

which, using (M4) (page 113), becomes

$$\left(\frac{\partial T}{\partial p} \right)_S = + \frac{T}{C_p} \left(\frac{\partial V}{\partial T} \right)_p \qquad (12.4)$$

This is one of the so-called adiabatic equations for a closed hydrostatic system. Similarly, using Maxwell relation (M3) (page 113) and the result

$$T\left(\frac{\partial S}{\partial T}\right)_V = C_V$$

Equation (12.3b) becomes

$$\left(\frac{\partial T}{\partial V}\right)_S = -\frac{T}{C_V}\left(\frac{\partial p}{\partial T}\right)_V \qquad (12.5)$$

Both Equation (12.4) and Equation (12.5) involve one principal heat capacity and quantities that may be obtained from the equation of state of the system.

Combining Equations (12.4) and (12.5) gives

$$\frac{C_p}{C_V}\left(\frac{\partial V}{\partial p}\right)_S = \left(\frac{\partial V}{\partial p}\right)_T$$

or, writing $\gamma = C_p/C_V$,

$$\gamma\left(\frac{\partial V}{\partial p}\right)_S = \left(\frac{\partial V}{\partial p}\right)_T \qquad (12.6)$$

This is a third adiabatic equation. It enables an equation for an isentropic process to be obtained from γ and the equation of state.

For a system consisting of n moles of molecules of an ideal gas, with equation of state $pV = nRT$, Equations (12.4), (12.5) and (12.6) become, respectively,

$$\left(\frac{\partial T}{\partial p}\right)_S = \frac{T\,nR}{C_p\,p} \qquad (12.7)$$

$$\left(\frac{\partial T}{\partial V}\right)_S = \frac{T\,nR}{C_V\,V} \qquad (12.8)$$

$$\gamma\left(\frac{\partial V}{\partial p}\right)_S = -\frac{V}{p} \qquad (12.9)$$

When C_p can be treated as a constant, integration of Equation (12.7) gives [1]

$$\ln T - \frac{nR}{C_p}\ln p = \text{constant} \qquad (12.10)$$

while Equation (12.8) gives

$$\ln T + \frac{nR}{C_V} \ln V = \text{constant} \tag{12.11}$$

when C_V is treated as a constant. Finally, when γ may be treated as a constant, the integration of Equation (12.9) gives

$$\ln V + \ln p = \text{constant or}$$
$$pV^\gamma = \text{constant} \tag{12.12}$$

Since, for an ideal gas,

$$C_p - C_p = nR$$

Equation (12.10) may be written

$$T^\gamma p^{(1-\gamma)} = \text{constant} \tag{12.13}$$

and Equation (12.11) may be written

$$T V^{(\gamma-1)} = \text{constant} \tag{12.14}$$

In a finite reversible process the work done on a closed hydrostatic system as its volume changes from V_i to V_f is

$$W = -\int_{V_i}^{V_f} p \, dV$$

When the system is an ideal gas and the process is isentropic,

$$p = kV^{-\gamma}$$

where k is a constant, provided that γ is a constant. Then

$$W = -\int_{V_i}^{V_f} k V^{-\gamma} \, dV$$

$$= -\frac{k}{1-\gamma} (V_f^{1-\gamma} - V_i^{1-\gamma})$$

or, since $k = p_i V_i^\gamma = p_f V_f^\gamma$,

$$W = \frac{1}{\gamma-1} (p_f V_f - p_i V_i) \tag{12.15}$$

12.1.2 The measurement of γ

When an ideal gas undergoes an isentropic process, the relationship between p, V and T depends on γ, as given by Equations (12.12) – (12.14). These equations should hold very closely for dilute real gases, and so the measurement of γ is of some importance. The earliest measurement of γ was made on air by Désormes and Clément and was published in 1819. Their apparatus is shown in Figure 12.1. The air is contained in a large vessel A fitted with a wide stopcock B and a manometer aa'. Initially the air has a temperature T_i equal to that of the surroundings and a pressure p_i slightly below that of the surroundings. When p_i has been recorded, the stopcock is opened, and air flows in from the surroundings and compresses the air initially in the vessel. This occurs rapidly, so that there is no time for a significant thermal interaction with the surroundings. When the pressure in the vessel has reached p_0, that of the surroundings, the stopcock is closed. During the work interaction the temperature of the gas in the vessel rises. Following the closure of the stopcock a thermal interaction with the surroundings takes place and the temperature of the air in the vessel falls to that of the surroundings while its volume remains constant. This causes the pressure of the air in the vessel to fall and, when equilibrium is established, its pressure is p_f.

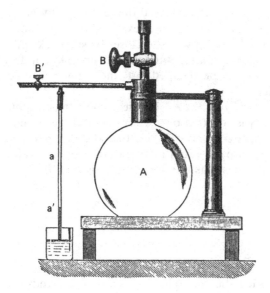

Figure 12.1 *The apparatus of Désormes and Clément for the measurement of the ratio of the principal specific heat capacities of a gas. This is Figure 135 on page 353 of* A Text-book of Heat: Part I, *by H. S. Allen and R. S. Maxwell, Macmillan, London (1959 reprint of corrected edition of 1944. First edition 1939)*

Consider the behaviour of n moles of molecules of air that are in the vessel and occupy a volume V_i before the stopcock is opened. This air constitutes the system. If the adiabatic compression is reversible and γ is constant over the range of temperature involved,

$$p_i V_i^\gamma = p_0 V_f^\gamma \qquad (12.16)$$

provided that the air may be treated as an ideal gas. V_f is the final volume of the system. For any reversible process undergone by an ideal gas

$$pV = nRT$$

and, since the initial and final temperatures of the air in the vessel are the same,

$$p_i V_i = p_f V_f \qquad (12.17)$$

Eliminating V_i and V_f from Equations (12.16) and (12.17) gives

$$p_i/p_0 = (p_i/p_f)^\gamma$$

or

$$\gamma = \frac{\ln p_i - \ln p_0}{\ln p_i - \ln p_f} \qquad (12.18)$$

Gay-Lussac and Welter improved the technique by making p_i slightly greater than p_0 and allowing the air to expand, but the major problem with the arrangement is knowing when to close the stopcock: pressure oscillations are set up as the air passes through the stopcock, but if sufficient time is allowed for these to be damped out, the expansion of the air is not closely adiabatic.

Measurements of γ for the noble gases and metallic vapours show that, at low pressures, γ is constant over a very wide range of temperature and has a value of approximately 5/3. For the so-called permanent diatomic gases γ is constant around room temperature, having a value of approximately 7/5, but decreases as the temperature is raised.

12.1.3 Heat engines

Thermodynamics may be said to have evolved from a study of heat engines — particularly, those designed by Newcomen and Watt in the eighteenth century [2]. However, because of its generality, thermodynamics is able to say very little about the detailed behaviour of real heat engines, although it does provide an aid to the understanding of the processes that occur and is able to place an upper limit on the thermal efficiency that may be obtained.

To discuss the performance of a real (cyclic) heat engine an idealised cycle is devised which is considered to be a reasonable representation of the real cycle. In this idealised cycle it is assumed that all processes are reversible and that the working substance is a simple pure substance. For the representation of the steam engine this latter condition is fulfilled in practice, but in idealisations of internal combustion engines air is usually taken as the working substance and the further assumption is usually made that air is an ideal gas. Ideal cycles using these assumptions are called *air-standard cycles*. In real internal combustion engines the working substance is a mixture of air and other gases, and the reactions that occur change the composition during the cycle. The main result that the idealised discussion can hope to achieve is to provide some upper bound for the efficiency of the real engine. As an example of a simple application of thermodynamics to heat engines, a brief discussion will be given of the air-standard petrol cycle.

The basic cycle of operations of spark-ignition engines, such as the petrol engine, consists of the following six stages:

1 A mixture of air and petrol vapour is drawn into the cylinder of the engine.
2 The mixture is rapidly compressed, a process which is approximately adiabatic, and a large temperature rise results.
3 A spark is produced which causes the mixture to explode, giving a sudden increase in temperature and pressure. This occurs at the end of the piston stroke, so that there is virtually no piston movement during the explosion.
4 The hot gas expands in an approximately adiabatic way, pushing back the piston and doing mechanical work (the power stroke).
5 At the end of the power stroke an outlet valve is opened and hot exhaust gases rush out to the atmosphere.
6 Finally, the piston forces out the remaining gas.

The idealised representation of the above cycle is the *air-standard Otto cycle*. In this cycle all the processes are assumed to be reversible, the approximately adiabatic processes are assumed to be strictly so, processes (3) and (5) are assumed to take place at constant volume, and the working substance (air) is treated as an ideal gas with constant heat capacities. The indicator diagram for this ideal cycle is shown in Figure 12.2. To change the temperature of the air at constant volume, stages 3 and 5 in Figure 12.2, the air must absorb heats Q_1 and Q_2, respectively. Then, using the notation of Figure 12.2, $Q_1 = mC_V(T_c - T_b)$ and $Q_2 = mC_V(T_a - T_d)$, where m is the mass of air in the cylinder. The efficiency η of the cycle is $-W/Q_1$, where W is the work done on the air per cycle and $W + Q_1 + Q_2 = 0$. Therefore,

$$\eta = \frac{Q_1 + Q_2}{Q_1} = 1 + \frac{Q_2}{Q_1} = 1 + \frac{(T_a - T_d)}{(T_c - T_b)} \qquad (12.19)$$

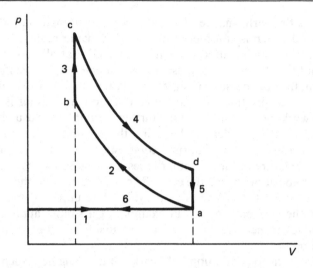

Figure 12.2 *The indicator diagram for the air-standard Otto cycle*

If air is considered to be an ideal gas,

$$T_b = T_a r^{\gamma-1} \text{ and } T_c = T_d r^{\gamma-1}$$

where r, equal to V_a/V_b, is known as the *compression ratio*. Then

$$\eta = 1 + \frac{(T_b r^{1-\gamma} - T_c r^{1-\gamma})}{T_c - T_b} = 1 - r^{1-\gamma} = 1 - \frac{1}{r^{\gamma-1}} \quad (12.20)$$

For modern cars r is about 10, so that taking γ as 1.4 gives a value of η of about 0.6. In practice η is about 0.3. The ideal Otto cycle is less efficient than a Carnot engine operating between the same temperature limits, since the heat transfer does not take place wholly at the maximum and minimum temperature. The value of η increases as r is increased, but for large values of r, the temperature after compression is high and may cause the fuel to self-ignite.

12.2 Perfectly elastic solids

The range of strains over which real solids deform elastically is very small, except for some polymeric materials, and the accompanying volume change is usually negligible. Under such conditions the displacement work done on the rod may be neglected. When conditions are restricted to those in which strains are very small and the behaviour of the material is perfectly elastic, the change in the internal energy U in an infinitesimal process is given by

$$dU = TdS + FdL \quad (12.21)$$

for a rod whose length is L under a tension F when the temperature is T and S is the entropy. It was shown in Section 5.2 that, when the process is reversible, so that $T dS$ is equal to the heat absorbed by, and $F dL$ is the work done on the rod, the change in length of a rod of cross-sectional area A is given by

$$dL = L\lambda dT + \frac{L}{AE} dF \tag{12.22}$$

where λ is the linear expansivity of the material and E its Young's modulus.

Equation (12.21) provides some interesting clues about the internal behaviour of elastic solids, despite its being a 'macroscopic' equation. Differentiating Equation (12.21) with respect to L, keeping T constant, gives

$$F = \left(\frac{\partial U}{\partial L}\right)_T - T\left(\frac{\partial S}{\partial L}\right)_T \tag{12.23}$$

This equation contains both U and S, neither of which can be measured directly. However, using the elastic solid equivalent of (M3) (page 113) [3],

$$\left(\frac{\partial S}{\partial L}\right)_T = -\left(\frac{\partial F}{\partial T}\right)_L \tag{12.24}$$

Equation (12.23) becomes

$$F = \left(\frac{\partial U}{\partial L}\right)_T + T\left(\frac{\partial F}{\partial T}\right)_L \tag{12.25}$$

This equation allows $(\partial U/\partial L)_T$ to be determined, since F, T and L can all be measured.

For simple crystalline solids, such as copper and rock salt, $T(\partial F/\partial T)_L$ is negative and negligible compared with the value of $(\partial U/\partial L)_T$. In contrast, for rubber $(\partial U/\partial L)_T$ is negligible compared with $T(\partial F/\partial T)_L$, which is positive.

This difference in behaviour is a consequence of the very different structures of these materials. In simple crystalline solids the applied tension changes the separation of the atoms or ions, the work of stretching going largely into increasing the internal energy of the material. However, the work done in stretching rubber at constant temperature goes mostly to decreasing the entropy and is transferred to the surroundings as heat. Rubber is a high polymer which, in the vulcanised condition, has some bonds, or cross-links, between adjacent long-chain molecules, giving a coarse three-dimensional network. The polymer chains are in a condition of continuous agitation and, in the absence of an applied tension, are coiled, folded or twisted in a more or less random manner. It is the small number of cross-links that makes the material a solid rather than a liquid: the cross-links allow for local changes in chain arrangements and yet maintain a coherent solid structure. When a tension is applied to a sample of vulcanised rubber, the chains slide past their neighbours, so that they are pulled out of their random arrangement and line up progressively along the direction of tension, an effect that is limited by the presence of the cross-links. It is this partial lining up of the polymer chains that accounts for the decrease in entropy.

When an elastic rod is deformed isentropically, it suffers, in general, a change in temperature given by $(\partial T/\partial L)_S$. None of the equations introduced so far in this section are very suitable for describing this temperature change but, as was found for closed hydrostatic systems, a reciprocity relation can be used to give a suitable equation.

Consider the reciprocity relation

$$\left(\frac{\partial T}{\partial L}\right)_S \left(\frac{\partial L}{\partial S}\right)_T \left(\frac{\partial S}{\partial T}\right)_L = -1$$

which may be written

$$\left(\frac{\partial T}{\partial L}\right)_S = -\left(\frac{\partial S}{\partial L}\right)_T \left(\frac{\partial T}{\partial S}\right)_L \tag{12.26}$$

By analogy with Equation (11.11), a heat capacity at constant length C_L can be defined by

$$C_L = T\left(\frac{\partial S}{\partial T}\right)_L \tag{12.27}$$

Using this equation for $(\partial T/\partial S)_L$ and Equation (12.24) for $(\partial S/\partial L)_T$ gives

$$\left(\frac{\partial T}{\partial L}\right)_S = \frac{T}{C_L}\left(\frac{\partial F}{\partial T}\right)_L \tag{12.28}$$

From Equation (12.22)

$$\left(\frac{\partial F}{\partial T}\right)_L = -AE\lambda$$

so that Equation (12.28) may be written in the more usual form

$$\left(\frac{\partial T}{\partial L}\right)_S = -\frac{T}{C_L}AE\lambda \tag{12.29}$$

When the change in length of the elastic rod takes place at constant temperature, there is an absorption of heat by the material. For a reversible change in length an expression for the heat absorbed, dQ, may be obtained from a suitable expression for TdS, since, in a reversible process, $dQ = TdS$. The entropy of an elastic rod for which volume changes are negligible may be written as a function of T and F:

$$S = S(T,F)$$

This gives

$$dS = \left(\frac{\partial S}{\partial T}\right)_F dT + \left(\frac{\partial S}{\partial F}\right)_T dF$$

so that

$$dQ = TdS = T\left(\frac{\partial S}{\partial T}\right)_F dT + T\left(\frac{\partial S}{\partial F}\right)_T dF \qquad (12.30)$$

$T\left(\frac{\partial S}{\partial T}\right)_F$ is the heat capacity under constant tension. When the process is isothermal dT is zero so that, using the Maxwell relation equivalent to (M1) (page 114),

$$\left(\frac{\partial S}{\partial F}\right)_T = \left(\frac{\partial L}{\partial T}\right)_F \qquad (12.31)$$

the expression for dQ, which may be written dQ_T, becomes

$$dQ_T = T\left(\frac{\partial L}{\partial T}\right)_F dF = T\,\lambda L dF \qquad (12.32)$$

or, using Equation (12.22) with T constant,

$$dQ_T = TAE\,\lambda dL \qquad (12.33)$$

Equation (12.30) can also be used to give $(\partial T/\partial L)_S$, without the need to use a reciprocity relation. This approach will be used for the other systems treated in this chapter.

12.2.1 The measurement of Young's modulus and linear expansivity

The simplest method for determining Young's modulus E for a material in the form of a rod or wire of length L is to suspend the rod vertically from a rigid support and measure the extension dL produced by an increase in the tension dF. If the area of cross-section A is also determined, E under isothermal conditions can be found, using the equation

$$E = \frac{L}{A}\left(\frac{\partial F}{\partial L}\right)_T \qquad (12.34)$$

Modern testing machines can examine specimens in the form of short cylinders to be tested in compression as well as long specimens for testing in tension. Further, the specimen is usually tested at a constant extension rate by moving one end of the specimen at constant velocity while the other remains fixed. The specimen then deforms to maintain the imposed strain rate and the force needed to produce this strain rate is measured by a suitable load cell.

Young's modulus is temperature-dependent; for simple crystalline materials E decreases as temperature increases. Values of E at room temperature for some common materials are given in Table 12.1. For the metals these values refer to measurements on isotropic (fine-grained) commercially pure materials.

A precision method for measuring the linear expansivity λ of a solid was developed by Fizeau in the 1860s. A block of material was prepared with two parallel faces, one of which was polished. This block was placed on a table and above it an optical flat was supported so that the surfaces of the optical flat made a very small angle with the polished surface of the crystal. When monochromatic light was directed normally onto the solid,

Table 12.1 Values of Young's modulus E at room temperature for some common materials. For metals the values of E refer to fine-grained samples

Material	E/GPa
Aluminium	70.3
Copper	129.8
Gold	78.0
Lead	16.1
Silver	82.7
Zinc	108.4
Mild steel	211.9
Nylon 66	1.2–2.9
Polypropylene	1.1–1.6
Polymethylmethacrylate	2.4–3.4

Values from Kaye and Laby (1973)

an interference pattern was produced by light reflected from the lower surface of the optical flat and from the upper, polished surface of the solid. The value of λ relative to the support could be determined from the movement of the fringe pattern with change in temperature. Repeating the measurements with the solid removed allowed the effects of the support to be determined.

Small displacements may be measured, using the change in capacitance of a parallel-plate capacitor. The expansion of the solid is transmitted to one plate of such a capacitor, while the other plate remains fixed. This small movement results in a change in capacitance, which may be measured with a bridge circuit, and this change may be related to the corresponding displacement of the solid. Again a correction is needed for the expansion of the apparatus.

For crystalline materials absolute values of λ may be obtained from X-ray determinations of the lattice parameter as a function of temperature. This approach has the advantage that it is not complicated by changes in the dimensions of the specimen produced by vacancy formation and the presence of impurities. However, the sensitivity of the technique is not as high as that of optical methods.

The cubic expansivity β of solids may be calculated from appropriate measurements of the linear expansivity. Consider a solid specimen in the form of a cuboid and having edge lengths in the three axial directions of L_1, L_2 and L_3, respectively. Then the volume V of the specimen is

$$V = L_1 L_2 L_3$$

so that

$$\left(\frac{\partial V}{\partial T}\right)_p = L_2 L_3 \left(\frac{\partial L_1}{\partial T}\right)_p + L_3 L_1 \left(\frac{\partial L_2}{\partial T}\right)_p + L_1 L_2 \left(\frac{\partial L_3}{\partial T}\right)_p$$

Now β is defined by

$$\beta = \frac{1}{V}\left(\frac{\partial V}{\partial T}\right)_p$$

and, therefore

$$\beta = \frac{1}{L_1}\left(\frac{\partial L_1}{\partial T}\right)_p + \frac{1}{L_2}\left(\frac{\partial L_2}{\partial T}\right)_p + \frac{1}{L_3}\left(\frac{\partial L_3}{\partial T}\right)_p = \lambda_1 + \lambda_2 + \lambda_3 \quad (12.35)$$

where λ_1, λ_2 and λ_3 are, respectively, the linear expansivities in the three axial directions.

To determine β for liquids it is usual to infer changes in the volume from changes in the density: for a liquid of density ρ the cubic expansivity is given by

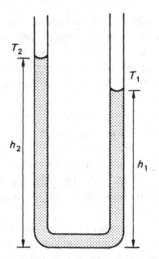

Figure 12.3   *The apparatus of Dulong and Petit for the measurement of the cubic expansivity of liquids*

$$\beta = -\frac{1}{\rho}\left(\frac{\partial \rho}{\partial T}\right)_p \qquad (12.36)$$

In the method introduced by Dulong and Petit two columns of the liquid, maintained at different temperatures, are allowed to balance, as in Figure 12.3. Then, if h_1 is the height of the liquid column at a temperature T_1, when the liquid density is ρ_1,

$$h_1\, \rho_1 = h_2\, \rho_2 \qquad (12.37)$$

This method does not require a knowledge of the expansivity of the container but, with the simple arrangement shown, the method is subject to a number of errors [4] which were eliminated by Regnault and others.

Values of λ and β for some common materials are given in Table 12.2. An interesting result, known as Gruneisen's law, is found to hold approximately for metals, particularly at higher temperatures: for a given metal the expansivity is proportional to the specific heat capacity at constant pressure. This result is illustrated for copper by the values in Table 12.3.

12.3 Liquid–vapour interfaces

Consider as the thermodynamic system a film of a pure liquid formed on a frame, as in Figure 5.4. The film is in equilibrium with the surrounding

Table 12.2 Values of linear expansivity λ and cubic expansivity β at room temperature for some common materials. For metals the values of λ refer to fine-grained samples

Material	$10^6\lambda/K^{-1}$	$10^5\beta/K^{-1}$
Aluminium	23.0	
Copper	16.7	
Gold	14	
Lead	29	
Silver	19	
Zinc (cast)	30	
Carbon steel	11	
Acetone		143
Carbon tetrachloride		122
Chloroform		127
Mercury		18.1
Turpentine		96
Water		21

Values from Kaye and Laby (1973).

Table 12.3 The Gruneisen law for copper

T/K	$C_{p,m}/J\ mol^{-1}\ K^{-1}$	$10^6\lambda/K^{-1}$	$10^6\lambda \div C_{p,m}/mol\ J^{-1}$
50	6.25	3.8	0.608
100	16.1	10.5	0.652
150	20.5	13.6	0.663
200	22.8	15.1	0.662
250	24.0	16.1	0.671
300	24.5	16.8	0.686
500	25.8	18.3	0.709
800	27.7	20.0	0.722
1200	30.2	23.4	0.775

Values from Roberts and Miller (1951).

vapour and it is assumed that, as the film changes its surface area, there is no net transfer of molecules between the film and the vapour. Then, provided that the volume of the liquid remains constant when the surface area changes [5], equilibrium states are represented by points in σ–T–A space and the internal energy U of the liquid when its temperature is T is given by

$$dU = TdS + \sigma dA \qquad (12.38)$$

where S is the entropy of the liquid, σ is the specific surface free energy and A is the surface area.

To discuss reversible isothermal and isentropic changes of the surface area of this system, it is necessary to obtain an expression for the change in entropy in any process. With the system of Figure 5.4 and the assumption that the volume is constant, two independent coordinates are needed to specify equilibrium states. The entropy may, therefore, be written as a function of two coordinates and it is convenient to choose A and T. Then

$$dS = \left(\frac{\partial S}{\partial A}\right)_T dA + \left(\frac{\partial S}{\partial T}\right)_A dT$$

or

$$TdS = T\left(\frac{\partial S}{\partial A}\right)_T dA + T\left(\frac{\partial S}{\partial T}\right)_A dT \tag{12.39}$$

When the process is reversible, $T(\partial S/\partial T)_A$ may be interpreted as the heat capacity C_A of the system when the surface area is held constant — that is,

$$C_A = T\left(\frac{\partial S}{\partial T}\right)_A \tag{12.40}$$

Also, for a reversible process, the other partial derivative may be transformed by means of a Maxwell relation appropriate to the system. The relation needed here is

$$-\left(\frac{\partial S}{\partial A}\right)_T = \left(\frac{\partial \sigma}{\partial T}\right)_A \tag{12.41}$$

which may be obtained from Equation (12.38) together with a reciprocity relation. Substituting for $(\partial S/\partial T)_A$ and $(\partial S/\partial A)_T$ into Equation (12.39) gives

$$TdS = C_A dT - T\left(\frac{\partial \sigma}{\partial T}\right)_A dA \tag{12.42}$$

Experimentally it is found that, except for very thin films, σ is a function of T only, and is independent of A. Therefore, Equation (12.42) may be written

$$TdS = C_A dT - T\frac{d\sigma}{dT} dA \tag{12.43}$$

When the liquid surface area is changed under reversible isothermal conditions, dT is zero; this is achieved by the liquid absorbing a quantity of heat dQ_T in an infinitesimal process, given by

$$dQ_T = TdS = -T\frac{d\sigma}{dT}dA \qquad (12.44)$$

For such a process Equation (12.38) becomes

$$dU_T = \left(\sigma - T\frac{d\sigma}{dT}\right)dA$$

or, for a finite process in which the surface area changes from A_i to A_f,

$$\left(\frac{U_f - U_i}{A_f - A_i}\right)_T = \sigma - T\frac{d\sigma}{dT} \qquad (12.45)$$

As the pressure and temperature of the bulk liquid have not altered in this process, the change in internal energy given by Equation (12.45) must be associated solely with the surface region. $[(U_f - U_i)/(A_f - A_i)]_T$ is the internal energy per unit area of the liquid surface and is called the *specific total surface energy*. Note that the specific surface free energy and the specific total surface energy are equal if σ is independent of T.

Values of σ may be obtained from the excess pressure in a bubble over that of the surroundings (see Exercise 7, in Chapter 5, but note that, if the bubble is formed in the bulk liquid, the value of the excess pressure is $2\sigma/r$).

Values of σ for some common pure liquids are given in Table 12.4. The value of σ decreases with increase in temperature for all liquids and becomes zero in the neighbourhood of the critical temperature.

Table 12.4 Values of the specific surface free energy σ at 20°C for some common liquids

Liquid	A/V	$\sigma/\text{mJ m}^{-2}$
Acetone	V	23.7
Carbon tetrachloride	V	27.0
Ethanol	V	22.8
Mercury	V	472
Methanol	A	22.6
Water	A	72.75

V = σ measured against own vapour; A = σ measured against air.
Values from Kaye and Laby (1973).

12.4 **Paramagnetic solids**

For a paramagnetic solid in which volume changes are negligible, equilibrium states are specified by the set of values of temperature T, applied flux density B_a (the B-field) and magnetic moment m. In Section 5.4 it was shown that the work that must be done on a specimen to increase the magnetic moment reversibly by an amount dm in a B-field of magnitude B_a is $B_a dm$ [6]. The change in the internal energy U of the specimen is then given by

$$dU = TdS + B_a dm \qquad (12.46)$$

where S is the entropy of the specimen.

With the conditions given, a paramagnetic solid has two degrees of freedom, so that the entropy may be written as a function of two coordinates. It is convenient to choose T and B_a, so that the entropy may be written

$$S = S(T, B_a)$$

Then, in an infinitesimal process,

$$TdS = T\left(\frac{\partial S}{\partial T}\right)_{B_a} dT + T\left(\frac{\partial S}{\partial B_a}\right)_T dB_a \qquad (12.47)$$

When the process is reversible, $T(\partial S/\partial T)_{B_a}$ is the heat capacity C_{B_a} of the system in a constant applied B-field, and $(\partial S/\partial B_a)_T$ may be transformed, using the Maxwell relation

$$\left(\frac{\partial S}{\partial B_a}\right)_T = \left(\frac{\partial m}{\partial T}\right)_{B_a} \qquad (12.48)$$

The derivation of Equation (12.48) is left as an exercise for the reader. Substituting for the two partial derivatives in Equation (12.47) gives

$$TdS = C_{B_a} dT + T\left(\frac{\partial m}{\partial T}\right)_{B_a} dB_a \qquad (12.49)$$

Now the magnetisation M of the specimen is given by $M = m/V$, where V is the volume of the specimen. Therefore, Equation (12.49) may be written

$$TdS = C_{B_a} dT + TV\left(\frac{\partial M}{\partial T}\right)_{B_a} dB_a \qquad (12.50)$$

In a reversible isothermal magnetisation of the specimen dT is zero and, as Equation (12.50) shows, such a change is accompanied by the specimen absorbing a quantity of heat dQ_T, given by

$$dQ_T = TdS_T = TV\left(\frac{\partial M}{\partial T}\right)_{B_a} dB_a \qquad (12.51)$$

The heat absorbed in such a process is zero when M is independent of temperature.

When the sample undergoes an isentropic process, dS is zero and there is an accompanying change in temperature dT_S, given by

$$\frac{dT_S}{T} = -\frac{V}{C_{B_a}}\left(\frac{\partial M}{\partial T}\right)_{B_a} dB_a \qquad (12.52)$$

If M is temperature-dependent, there is a change in the temperature of the specimen when the magnetisation is changed isentropically. This is known as the *magnetocaloric effect*.

For a linear, isotropic, homogeneous paramagnetic material at constant temperature, $\mu_0 M/B_a$ is a scalar, independent of the magnitude of B_a and is known as the magnetic (volume) susceptibility χ_m of the material. Here μ_0 is the permeability of a vacuum. The advantage of using χ_m is that it is straightforwardly determined from measurements of the force on a magnetic material in a non-uniform magnetic field. In Gouy's method the specimen is in the form of a long rod of uniform cross-section A. The specimen is suspended with its axis vertical so that the lower end is in the field at the mid-position between the flat pole tips of a magnet, where B_a has the value B_{a0} and the upper end is in a field-free zone. The total downward force F on the specimen is then

$$F = \frac{(\chi_m - \chi'_m)B_{a0}^2}{2\mu_0} \qquad (12.53)$$

where χ'_m is the susceptibility of the surrounding medium, usually air [7]. Values of χ_m for some common paramagnetic materials are given in Table 12.5.

Table 12.5 Values of the susceptibility χ_m at 20°C for some paramagnetic materials

Material	χ_m
Aluminium	2.08×10^{-5}
Caesium	5.15×10^{-6}
Iridium	3.75×10^{-5}
Magnesium	1.18×10^{-5}
Sodium	8.5×10^{-6}
Uranium	4.11×10^{-4}

Values from Nordling and Österman (1980).

To a good approximation paramagnetic solids obey the Curie–Weiss law

$$\chi_m = \frac{a}{T - T_c} \tag{12.54}$$

where a is a constant and T_c is a temperature, characteristic of the material, known as the *Curie temperature*. For many paramagnetics T_c is a fraction of a kelvin, so that, except at the lowest temperatures, an adequate representation of the magnetic behaviour is given by the Curie law

$$\chi_m = \frac{a}{T} \tag{12.55}$$

Knowing the temperature dependence of χ_m allows Equations (12.51) and (12.52) to be integrated. Since $\chi_m = \mu_0 M / B_a$, Equation (12.51) becomes

$$dQ_T = \frac{T V B_a}{\mu_0} \left(\frac{\partial \chi_m}{\partial T} \right)_{B_a} dB_a$$

and, for a material that obeys Curie's law,

$$\left(\frac{\partial \chi_m}{\partial T} \right)_{B_a} = -\frac{a}{T^2}$$

Therefore,

$$dQ_T = -\frac{V B_a \, a}{\mu_0 T} dB_a$$

In a finite reversible isothermal magnetisation, in which the applied B-field changes from B_{ai} to B_{af}, the heat absorbed is given by

$$Q_T = -\frac{V a}{2\mu_0 T} (B_{af}^2 - B_{ai}^2) \tag{12.56}$$

When B_{af} is greater than B_{ai}, the negative sign indicates that the heat flow is from the specimen to the surroundings.

In terms of the susceptibility χ_m, the change in temperature of a paramagnetic specimen in an isentropic magnetisation is

$$\frac{dT_S}{T} = -\frac{V B_a}{C_{B_a} \mu_0} \left(\frac{\partial \chi_m}{\partial T} \right)_{B_a} dB_a \tag{12.57}$$

To integrate Equation (12.57) it is necessary to know how χ_m depends on T when B_a is held constant and also how C_{B_a} depends on B_a when T is held constant. As has already been seen, for a material obeying Curie's law,

$$\left(\frac{\partial \chi_m}{\partial T}\right)_{B_a} = -\frac{a}{T^2}$$

Now C_{B_a} is $T(\partial S/\partial T)_{B_a}$ and can, therefore, only be a function of B_a and T. Then

$$\left(\frac{\partial C_{B_a}}{\partial B_a}\right)_T = T\frac{\partial}{\partial B_a}\Bigg)_T\left(\frac{\partial S}{\partial T}\right)_{B_a} = T\frac{\partial}{\partial T}\Bigg)_{B_a}\left(\frac{\partial S}{\partial B_a}\right)_T$$

and, using Equation (12.48), this becomes

$$\left(\frac{\partial C_{B_a}}{\partial B_a}\right)_T = T\frac{\partial}{\partial T}\Bigg)_{B_a}\left(\frac{\partial m}{\partial T}\right)_{B_a} = T\left(\frac{\partial^2 m}{\partial T^2}\right)_{B_a} \tag{12.58}$$

Equation (12.58) may, in turn, be rewritten, using V and χ_m, to give

$$\left(\frac{\partial C_{B_a}}{\partial B_a}\right)_T = \frac{T V B_a}{\mu_0}\left(\frac{\partial^2 \chi_m}{\partial T^2}\right)_{B_a} \tag{12.59}$$

For a material obeying Curie's law

$$\left(\frac{\partial^2 \chi_m}{\partial T^2}\right)_{B_a} = \frac{2a}{T^3}$$

so that, for such material,

$$C_{B_a}(B_a,T) - C_{B_a}(0,T) = \frac{a V B_a^2}{\mu_0 T^2}$$

Further, it is found that, at temperatures well above T_c, $C_{B_a}(0,T)$ is given by bV/T^2, where b is a constant. Therefore,

$$C_{B_a}(B_a,T) = \frac{a V B_a^2}{\mu_0 T^2} + \frac{b V}{T^2} = \frac{V}{T^2}\left(\frac{aB_a^2 + \mu_0 b}{\mu_0}\right)$$

so that, for a material that obeys Curie's law, Equation (12.57) becomes

$$\frac{dT_S}{T} = \frac{a B_a \, dB_a}{(aB_a^2 + \mu_0 b)} \tag{12.60}$$

Integrating Equation (12.60) gives

$$\ln T = \tfrac{1}{2} \ln (aB_a^2 + \mu_0 b) + \text{constant}$$

so that, if changing the applied B-field from B_{ai} to B_{af} under isentropic conditions changes the temperature of the paramagnetic solid from T_i to T_f,

$$\ln\left(\frac{T_f}{T_i}\right)_S = \tfrac{1}{2} \ln\left(\frac{aB_{af}^2 + \mu_0 b}{aB_{ai}^2 + \mu_0 b}\right) \tag{12.61}$$

If B_{af} is less than B_{ai} (i.e. the material is demagnetised isentropically), there is a fall in temperature. In particular, when B_{af} is zero,

$$\left(\frac{T_f}{T_i}\right)_S = \sqrt{\left(\frac{b}{b + aB_{ai}^2/\mu_0}\right)} \tag{12.62}$$

The above effect is the basis of the process known as *magnetic cooling* or *adiabatic demagnetisation*. In this process a paramagnetic salt with a very low value of T_c has its temperature reduced to about 1 K by being placed in thermal contact (by means of a gas known as the *exchange gas*) with a bath of liquid helium boiling under reduced pressure (see subsection 16.4.1). The sample is then magnetised reversibly and isothermally in a very strong magnetic field (≈ 5 T), a process which aligns the magnetic dipoles of the salt. This represents a decrease in the magnetic entropy which is then transferred to the lattice, the energy dissipated in the process being transmitted to the helium bath, through the exchange gas, as heat (Equation 12.51). The salt is then thermally isolated by the removal of the exchange gas and the applied magnetic field is reduced to zero isentropically. This produces a misorientation of the magnetic dipoles and is done at the expense of the lattice entropy. Consequently, the temperature of the salt falls (Equation 12.52).

12.5 Reversible voltaic cells

A cell that is operating in a thermodynamically reversible manner does not generate gas and has only an infinitesimal current flowing in the circuit. For such a cell it was shown in Section 5.5 that the work done on the cell by an external source when a charge dZ is moved from the negative electrode to the positive electrode in the external circuit, under a potential difference

E, is $E\mathrm{d}Z$. The change in internal energy of the cell resulting from this process is given by

$$\mathrm{d}U = T\mathrm{d}S + E\mathrm{d}Z \tag{12.63}$$

where S is the entropy of the cell.

A reversible voltaic cell has two degrees of freedom, so that the entropy may be written as a function of any two independent coordinates. It is convenient to choose T and Z and then write

$$S = S(T,Z)$$

From this equation it follows that, in any infinitesimal process,

$$\mathrm{d}S = \left(\frac{\partial S}{\partial T}\right)_Z \mathrm{d}T + \left(\frac{\partial S}{\partial Z}\right)_T \mathrm{d}Z$$

or

$$T\mathrm{d}S = T\left(\frac{\partial S}{\partial T}\right)_Z \mathrm{d}T + T\left(\frac{\partial S}{\partial Z}\right)_T \mathrm{d}Z \tag{12.64}$$

When the process is reversible, $T(\partial S/\partial T)_Z$ is the heat capacity at constant charge, denoted by C_Z. The derivative $(\partial S/\partial Z)_T$ can be replaced by one containing only primitive coordinates by using a Maxwell relation. For the voltaic cell the appropriate relations may be obtained from those for the closed hydrostatic system, replacing p by $-E$ and V by Z. The required Maxwell relation is

$$\left(\frac{\partial S}{\partial Z}\right)_T = -\left(\frac{\partial E}{\partial T}\right)_Z$$

Equation (12.64) may then be written

$$T\mathrm{d}S = C_Z\mathrm{d}T - T\left(\frac{\partial E}{\partial T}\right)_Z \mathrm{d}Z \tag{12.65}$$

and Equation (12.63) becomes

$$\mathrm{d}U = C_Z\mathrm{d}T + \left[E - T\left(\frac{\partial E}{\partial T}\right)_Z\right]\mathrm{d}Z \tag{12.66}$$

For a cell in which the electrolytes are saturated solutions E is a function of T only so that $(\partial E/\partial T)_Z$ may be written as the total derivative $\mathrm{d}E/\mathrm{d}T$. The expression for the entropy then becomes

$$dS = C_Z \frac{dT}{T} - \frac{dE}{dT} dZ \qquad (12.67)$$

In a reversible isothermal process dT is zero and the transfer of charge in the external circuit is accompanied by the absorption of heat by the cell. For an infinitesimal process the amount of this heat is dQ_T, given by

$$dQ_T = TdS_T = -T\frac{dE}{dT} dZ \qquad (12.68)$$

The corresponding change in the internal energy is given by

$$dU_T = \left(E - T\frac{dE}{dT} \right) dZ \qquad (12.69)$$

and, in a finite process, this becomes

$$\Delta U_T = \left(E - T\frac{dE}{dT} \right) \Delta Z \qquad (12.70)$$

The movement of charge in the external circuit is a consequence of chemical reactions taking place within the cell. When 1 mol of materials of the same valence n react as a result of a charge being drawn slowly from the cell, the amount of this charge is given by

$$\Delta Z_m = -nF \qquad (12.71)$$

where F is Faraday's constant. The heat absorbed by the cell during this process is given by

$$Q_T = nFT\frac{dE}{dT} \qquad (12.72)$$

and the change in the internal energy of the cell is given by

$$\Delta U_T = -nF\left(E - T\frac{dE}{dT} \right) \qquad (12.73)$$

Now, when the external pressure is constant and there is no change in the volume of the system,

$$\Delta U = \Delta H$$

where H is the enthalpy of the system. Equation (12.73) may, therefore, be written

$$\Delta H_T = - nF\left(E - T\frac{dE}{dT} \right) \tag{12.74}$$

where ΔH_T is the energy absorbed by the system when 1 mol of each constituent disappears under reversible isothermal conditions. It is known as the *heat of reaction*. The importance of this relation is that it allows thermal quantities to be determined by electrical measurements; it is only necessary to determine E as a function of T.

In the Daniell cell (see Section 5.5) the reaction within the cell is

$$Zn + Cu\,SO_4 \rightleftharpoons Cu + Zn\,SO_4$$

proceeding in the direction of the upper arrow when current flows in the external circuit from the copper to the zinc electrode. For this cell n is 2 and, at 273 K, E is 1.0934 V and dE/dT is -4.53×10^{-4} V K^{-1}. Equation (12.74) then gives a value of -235 kJ mol^{-1} for ΔH_T.

■ EXAMPLE

Q. A certain mass of an ideal gas is used as the working substance in a Carnot engine that is operating between reservoirs at 500 K and 200 K, respectively. When the gas is in contact with the high-temperature reservoir it expands from a pressure of 1.1×10^7 Pa and a volume of 1.2×10^{-3} m^3 to a pressure of 4.0×10^6 Pa.

(a) Calculate the limits of pressure and volume between which the working substance operates at the temperature of the low-temperature reservoir.
(b) Calculate the heat absorbed by the working substance from each reservoir during one cycle.

Take the ratio of the principal heat capacities γ of the gas to be independent of temperature and equal to 1.50.

A. Let the cycle of operations be represented on a pressure p–volume V indicator diagram, as in Figure 12.4.

Using an obvious notation, for the reversible isothermal expansion between states 1 and 2,

$$p_1 V_1 = p_2 V_2 \quad \text{(Boyle's law)}$$

Therefore,

$$V_2 = \frac{p_1 V_1}{p_2} = \frac{1.1 \times 1.2 \times 10^{-2}}{4} = 3.3 \times 10^{-3} \text{ m}^3$$

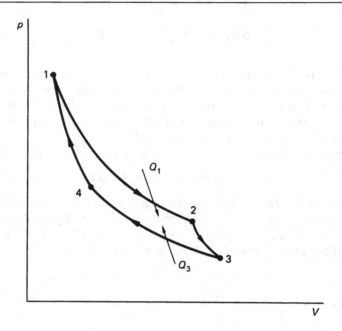

Figure 12.4 *Indicator diagram for an ideal gas performing a Carnot cycle*

For the isentropic expansion between states 2 and 3

$$T_2 V_2{}^{\gamma-1} = T_3 V_3{}^{\gamma-1} \quad \text{(adiabatic equation)}$$

Therefore,

$$V_3 = V_2\left(\frac{T_2}{T_3}\right)^{1/\gamma-1} = 7.3 \times 10^{-3}\left(\frac{500}{200}\right)^2 = 20.6 \times 10^{-3} \text{ m}^3$$

Also,

$$\frac{p_2 V_2}{T_2} = \frac{p_3 V_3}{T_3} \quad \text{(ideal gas equation)}$$

so that

$$p_3 = \frac{p_2 V_2}{T_2} \times \frac{T_3}{V_3} = \frac{4 \times 10^6 \times 3.3 \times 10^{-3} \times 200}{500 \times 20.6 \times 10^{-3}}$$

$$= 0.26 \times 10^6 \text{ Pa}$$

Further, for the isentropic compression between states 4 and 1,

$$T_1 V_1{}^{\gamma-1} = T_4 V_4{}^{\gamma-1}$$

giving

$$V_4 = V_1\left(\frac{T_1}{T_4}\right)^{1/\gamma-1} = 1.2 \times 10^{-3}\left(\frac{500}{200}\right)^2 = 7.5 \times 10^{-3} \text{ m}^3$$

For the reversible isothermal compression between states 3 and 4,

$$p_3V_3 = p_4V_4 \quad \text{(Boyle's law)}$$

so that

$$p_4 = \frac{p_3V_3}{V_4} = \frac{0.26 \times 10^6 \times 20.6 \times 10^{-3}}{7.5 \times 10^{-3}} = 0.71 \times 10^6 \text{ Pa}$$

Therefore, the limits of pressure and volume between which the gas operates at 200 K are

pressure : 0.26×10^6 Pa to 0.71×10^6 Pa
volume : 20.6×10^{-3} m^3 to 7.5×10^{-3} m^3

When an ideal gas undergoes a reversible isothermal process, the change in internal energy is zero, so that, from the first law, in an infinitesimal process $dQ = -dW$, where dQ is the heat absorbed by the gas and dW is the work done on the gas. In a finite process, therefore, the heat absorbed is Q, where

$$Q = \int dQ = -\int dW = -\int (-p\,dV)$$

Assume that the gas consists of n moles of molecules. Then, if V_m is the molar volume and R is the gas constant,

$$pV_m = RT = \frac{pV}{n}$$

so that

$$p_1V_1 = nRT_1$$

giving

$$nR = \frac{1.1 \times 10^7 \times 1.2 \times 10^{-3}}{500} = 26.4 \text{ J K}^{-3}$$

During the reversible isothermal process at 500 K

$$p = 26.4 \times \frac{500}{V} = \frac{1.32 \times 10^4}{V}$$

and

$$Q_1 = \int_{1.2\times10^{-3}}^{3.3\times10^{-3}} 1.32 \times 10^4 \frac{dV}{V} = 1.32 \times 10^4 \ln\left(\frac{3.3}{1.2}\right) = 1.34 \times 10^4 \text{ J}$$

The positive sign of Q_1 indicates that this heat transfer is from the reservoir to the gas. Now

$$\frac{Q_1}{Q_3} = -\frac{T_1}{T_3} \quad \text{(definition of thermodynamic temperature)}$$

Therefore,

$$Q_3 = -Q_1 \frac{T_3}{T_1} = -1.34 \times 10^4 \times \frac{200}{500} = -0.54 \times 10^4 \text{ J}$$

The negative sign of Q_3 indicates that the direction of heat flow is from the gas to the reservoir.

Appendix: Early heat engines

The first practical stationary steam engine was built by Thomas Newcomen in about 1712. Earlier, in 1698, Thomas Savery had obtained a patent for a device for ' . . . Raiseing of Water . . . by the Impellant Force of Fire', but this device had no moving parts and was not really a heat engine. In Savery's device steam from a boiler was admitted to a closed vessel, where it was condensed by cold water poured on the outside of the vessel. A partial vacuum was thereby produced within the vessel and atmospheric pressure forced water through a non-return valve in the bottom of the vessel. Steam was then admitted again to the vessel and this forced the water out of the vessel through a second non-return valve and along the delivery pipe, leaving a steam-filled vessel in which a partial vacuum could be produced by condensation.

Newcomen's engine used a piston moving in a vertical cylinder closed at its lower end only. The piston was connected to one end of a pivoted beam (see Figure 12.5), the other end of which was connected to a pump rod which actuated a pump.

To operate the engine, steam at a pressure slightly above atmospheric pressure was introduced into the cylinder when the piston was at its highest position and a partial vacuum was then produced by squirting water into the cylinder to condense the steam. The piston was then forced downwards by atmospheric pressure and the pump rod was raised. Steam was then admitted to the cylinder and the piston was raised by the weight of the pump rod. Any air or water in the cylinder was blown out through sealed

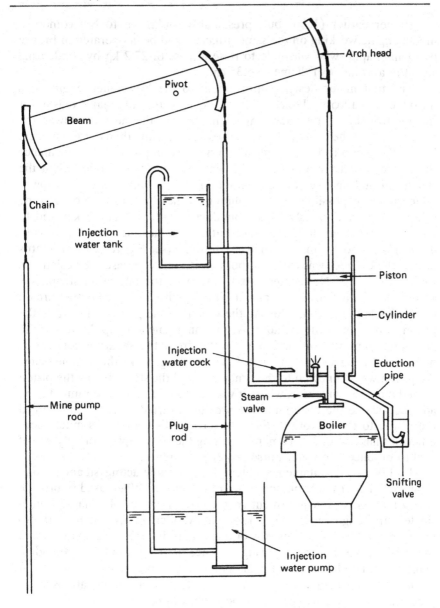

Figure 12.5 *A schematic representation of Newcomen's steam engine*

non-return valves during this operation. This was a reliable engine design, the major engineering problem being getting a good fit between piston and cylinder wall. The thermal efficiency of the Newcomen engine was, however, less than 1%, largely because the temperature of the cylinder had to be raised and lowered again during each cycle.

Rather earlier (1690) but, presumably, unknown to Newcomen, a model engine working on the same principle had been operated in France by Denis Papin, who was able to raise a mass of 27.2 kg by condensing steam in a cylinder of diameter 6.35 cm.

The first major design improvement to the Newcomen engine was introduced in 1765 by James Watt. He realised that the major reason for the inefficiency of the Newcomen engine was the need to lower the temperature of the cylinder to condense the steam and then to raise it again before the pump rod could raise the piston. To improve the performance the cylinder walls should be at the temperature of the boiler when the steam entered and yet the condensation should take place at as low a temperature as possible. Watt's solution to the problem was to keep the cylinder hot continuously by surrounding it with a steam-jacket and to condense the steam in a separate, but communicating, vessel (the condenser) kept cool by immersion in a water-bath (see Figure 12.6). In Watt's engine the cylinder was closed at both ends, but otherwise the design was similar to that of Newcomen. At the start of the operation steam was admitted to both sides of the piston. This equalised the steam pressure on both sides of the piston, allowing the weight of the pump rod to raise the piston to the top of the cylinder. At the same time a pump, known as the air pump, removed air and waste water from the separate condenser, maintaining a partial vacuum in it. In the working stroke the volume below the piston was connected to the condenser and the steam below the piston flowed into the condenser, where it was condensed. Steam continued to be admitted above the piston and the pressure exerted by this steam forced the piston to the bottom of the cylinder. Watt's design was much more efficient than that of Newcomen, requiring the consumption of only 9 lb of coal to produce one water horse power per hour.

In 1782 Watt obtained a patent for a double-acting steam engine: steam acted alternately on both sides of the piston. When used to provide rotary motion, this arrangement gave twice the power obtainable from a single-acting engine with the same size of cylinder. In his patent Watt also included the idea of expansive working: admitting steam for only part of the working stroke and allowing the steam to do work by expanding approximately adiabatically for the remainder of the stroke. However, at the low steam pressures used by Watt ($\approx 70 \times 10^3$ Pa above atmospheric pressure) expansive working gave no benefit in practice.

A further advance was introduced by Richard Trevithick, who, in 1797, built a stationary, high-pressure, non-condensing engine that did not infringe Watt's patent. Early models used steam pressures of about 360×10^3 Pa above atmospheric pressure, but one was built using steam at three times this pressure. In 1804 Trevithick built a steam-driven locomotive, again using a high-pressure, non-condensing design, which was driven by friction through its wheels and which blew its exhausted steam through

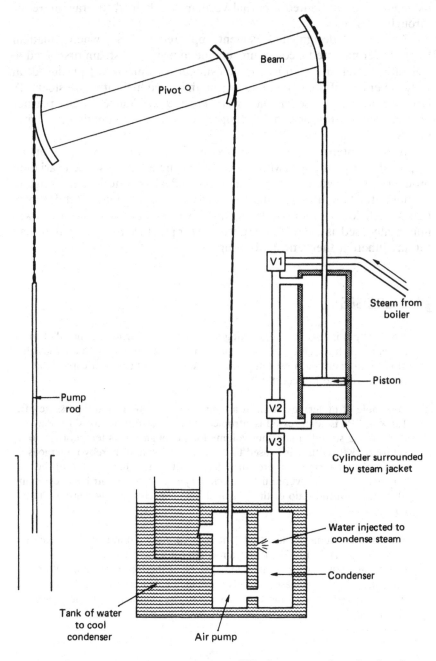

Figure 12.6 *A schematic representation of Watt's steam engine, showing the separate condenser. V1, Steam valve; V2, equilibrium valve; V3, exhaust valve*

its chimney. This created a partial vacuum and helped to draw more air through the fire.

A different design improvement appeared in 1781, when Jonathan Hornblower patented a compound engine in which the steam first acted in a small cylinder and then in a large cylinder. Arthur Woolf produced, in 1804, a version of the compound engine that used high-pressure steam. A large number of these engines were erected in France by his partner Humphrey Edwards, and so this design would have been familiar to Carnot (see Section 7.3).

It is of interest that all the above designs were developed on an empirical basis, although, with hindsight, it can be seen how each variation brought the design closer to that of the ideal thermodynamic engine. Carnot's study did little to enhance practical engine design, but it did show that the efficiency could not be increased without limit and that steam was universally used as the working substance for practical reasons, rather than for any inherent theoretical advantage.

■ EXERCISES

1　An ideal gas is used as the working substance in a Carnot engine. Plot the states through which the working substance passes in one cycle on a graph of internal energy against temperature. Assume that the heat capacity at constant volume is a constant.

2　One mole of molecules of air at a temperature of 300 K and a pressure of 1.0×10^5 Pa is compressed isentropically to a pressure of 2.0×10^5 Pa. After being stored at constant volume long enough for its temperature to return to 300 K, the air is used to drive an engine and thereby undergoes an isentropic expansion to a pressure of 1.0×10^5 Pa. What fraction of the original work is recovered if the machinery is perfect and air behaves as an ideal gas with the ratio of the principal heat capacities being constant and equal to 1.40?

3　A vessel of volume V has fitted to it an accurately ground vertical tube of cross-section A and is filled with air at atmospheric pressure p_0. A well-fitting ball-bearing of mass m is placed at the top of the tube. When the ball-bearing is released, it falls a distance L before coming instantaneously to rest and then starting to rise. Show that, if the process is isentropic and air can be treated as an ideal gas with constant heat capacities,

$$\gamma = \frac{2mgV}{p_0 A^2 L}$$

where γ is the ratio between the principal heat capacities and g is the acceleration of free fall.

4   Calculate the change in temperature of a nickel wire of diameter 2.0 mm as a result of the sudden application of a load of 20 kg. The initial temperature of the wire is 20 °C and nickel has the following properties:

specific heat capacity at constant length $= 0.44 \times 10^3$ J kg^{-1} K^{-1};
linear expansivity $= 3.0 \times 10^{-6}$ K^{-1};
density at 20 °C $= 8.8 \times 10^3$ kg m^{-3}.

Assume that the acceleration of free fall is 9.81 m s^{-2}.

5   Two similar rods, each of length L and the same area of cross-section, are made of different perfectly elastic materials, having Young's moduli E_1 and E_2 and linear expansivities λ_1 and λ_2, respectively. The rods are fixed end-to-end and the combined rod is rigidly clamped at its ends. When the temperature of the combined rod is raised by ΔT, show that the distance x that the junction moves from its original position is given by

$$x = \frac{L\, \Delta T (E_1\lambda_1 - E_2\lambda_2)}{E_1 + E_2}$$

6   A certain cloud of water vapour consists of droplets of diameter 90 nm. When this cloud condenses, drops of water of mass 1.0 g are produced. Calculate the change in temperature of the water as a result of the coalescence if the initial temperature of the drops is 323 K, assuming that the process is isentropic. The value of the specific surface free energy of water is 75.5 mJ m^{-2} at 273 K and 51.5 mJ m^{-2} at 373 K and, in this temperature range, may be considered to vary linearly with temperature. The specific heat capacity at constant surface area may be assumed to be the same as that at constant pressure and for water at 323 K is 4.18×10^3 J kg^{-1} K^{-1}.

7   Derive an expression for the change in entropy of a sample of a paramagnetic material, the susceptibility of which obeys the Curie–Weiss law, when it is magnetised reversibly and isothermally by the applied B-field changing from B_{ai} to B_{af}.

8   A certain cell has electrodes of univalent mercury. If the electromotive force E in volts is given by

$$E = 0.046 + 0.000\,34\,(T-300)$$

calculate the heat of reaction at 350 K. Faraday's constant is 9.648×10^4 C mol^{-1}.

9   A power plant consists of a boiler where steam is generated to drive a turbine that is connected to an electrical generator. After passing through the turbine the steam enters a condenser, and then a pump, driven by the turbine, returns the condensed steam to the boiler. Assume that the processes occurring in the boiler and condenser are approximately isothermal and that those occurring in the turbine and pump are approximately adiabatic. Take the temperature of the boiler to be 825 K and that of the condenser to be 275 K.

(a) Calculate the theoretical efficiency of the power plant, assuming that the cycle of operations is a Carnot cycle.
(b) In practice, the efficiency of this sort of power plant is about 0.3. Use this value to calculate the rate of heat transfer to the boiler to give a power output of 1.2 MW.
(c) Using the value obtained in (b), calculate the rate at which heat is rejected to the condenser.
(d) This rejected heat raises the temperature of water circulating around the condenser. If the maximum permitted temperature rise of this water is 5.0 K, calculate the minimum water circulation rate needed when the power output is 1.2 MW. The specific heat capacity of water at constant pressure is 4.2×10^3 J kg^{-1} K^{-1} and may be taken as constant.

This analysis assumes that the ideal cycle on which the real cycle is based is essentially a Carnot cycle. However, the Carnot engine with steam is impractical, as it is difficult to condense only partly a wet vapour and impossible to condense a wet vapour to liquid efficiently. In practice, a cycle based on an ideal cycle known as the Rankine cycle is used.

Notes

1 The constant in Equations (12.10)–(12.14) is really a function of the entropy and is a constant only in an isentropic process. The constant has a different value in each of Equations (12.10)–(12.14).

2 A brief description of the development of the steam engine in the eighteenth and early nineteenth centuries is given in the appendix to this chapter.

3 This result may be obtained by differentiating Equation (12.21) twice and then using a reciprocity relation. Alternatively, comparing Equations (12.21) and (12.1) shows that, to obtain the equivalent Maxwell relations for an elastic solid, it is only necessary to take the appropriate equation for a closed hydrostatic system and replace p by $-F$ and V by L.

4 For example, the two liquid surfaces are at different temperatures, so that their respective elevations because of surface effects will be different.

5 This assumption cannot be strictly true. The average volume occupied by a molecule in the surface region of a pure liquid must be somewhat greater than that occupied by a molecule in the interior if the liquid is to exhibit a surface tension.

6   As in Section 5.4, it is assumed here that the material is isotropic and homogeneous. Then all vector quantities are parallel and may be treated as scalars. In real paramagnetics the magnetisation is small and end effects (demagnetisation) may be neglected, whatever the shape of the sample.

7   See, for example, *Electricity and Magnetism,* 3rd edn, by W. J. Duffin (London: McGraw-Hill, 1980).

13

Equations of state

It was shown in Chapter 11 that both the internal energy and the entropy of a closed hydrostatic system can be determined completely when the equation of state and one principal heat capacity are known. This knowledge then gives complete information about the thermodynamic behaviour of the system. The measurement of heat capacities was discussed in Chapter 11, where it was shown that, for a complete knowledge of both principal heat capacities, it is only necessary to measure the heat capacity at constant pressure for one state on each isotherm, provided that the equation of state is known.

The equation of state of a system is the equation of the surface representing the totality of equilibrium states of the system. These equilibrium states are represented by sets of values of the appropriate coordinates — for example, pressure p, volume V and temperature T for a closed hydrostatic system.

To develop analytical equations of state it is necessary to examine in detail the form of the appropriate representative surface. This is done for the closed hydrostatic system composed of a pure substance in Section 13.1. In Section 13.2 some general principles are set out for developing equations of state on a sound thermodynamical basis and these are then applied to real gases.

13.1 Properties of pure substances

Pure substances constitute the simplest class of closed hydrostatic system. Each equilibrium state of such a system is represented by a point in p–V–T space, the totality of points giving a characteristic surface with the features shown schematically in Figure 13.1, which is for a pure substance that expands on melting. The dashed lines of Figure 13.1 separate the surface into regions in which different phases exist, where s denotes the solid

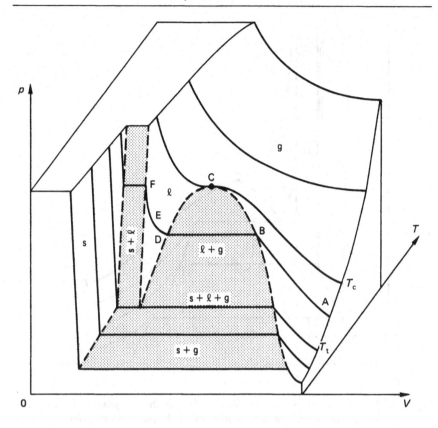

Figure 13.1 *The p–V–T surface for a pure substance that expands on melting. The full lines are isotherms; the dotted lines mark the phase boundaries. Mixed phases coexist in equilibrium in the shaded regions*

phase, l the liquid phase and g the gaseous phase. Several isotherms are shown as full lines.

Since three-dimensional surfaces are difficult to visualise, use is often made of projections. The projections commonly used with closed hydrostatic systems are those in the *p–V* plane, shown in Figure 13.2, and in the *p–T* plane, shown in Figure 13.3. Isotherms are usually plotted in the *p–V* projection and phase boundaries in the *p–T* projection.

For temperatures above T_c, which is known as the *critical temperature* of the substance, the substance is in the gaseous phase, whatever the value of the pressure. At temperatures T_c and below, it is always possible to produce a condensed phase of the substance if the pressure is sufficiently high. For this reason the gaseous phase is often termed 'vapour' when $T < T_c$, the term 'gas' being used when $T > T_c$.

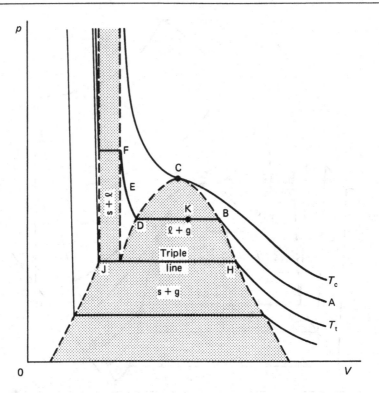

Figure 13.2 *The projection of the p–V–T surface in the p–V plane. The full lines are isotherms and the dashed lines mark the phase boundaries*

When the gas phase at the temperature T_t is compressed, condensation to the solid and liquid phases occurs simultaneously. T_t is known as the *triple-point temperature* and in the p–T diagram the three phases that coexist in equilibrium are represented by a point.

When condensation from the gaseous phase occurs at temperatures T in the range $T_t < T < T_c$, the condensation is to the liquid phase, while for $T < T_t$ condensation of the vapour is directly to the solid phase [1].

Consider the behaviour of the substance in the equilibrium states represented by the isotherm in ABDEF in Figure 13.2. The substance is in the gaseous phase at A and remains a gas as the volume is decreased until the state B is reached. Condensation to the liquid phase starts at B and continues at constant pressure as the volume is decreased, until at D all the substance is in the liquid phase. Along BD the system consists of liquid characterised by the coordinates of the point D and vapour characterised by the coordinates of the point B. The ratio between the mass of vapour m_g and the mass of liquid m_l at any point K is given by $m_l(BK) = m_g(KD)$ (see Exercise 8). This constant pressure, which is the maximum equilibrium pressure that the vapour is able to exert at that temperature, is known as

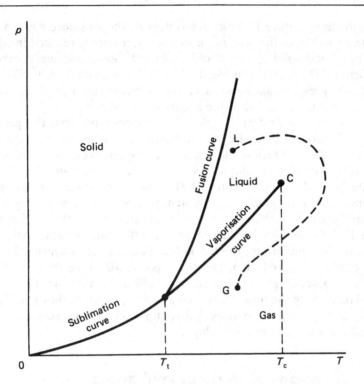

Figure 13.3 *The projection of the* p–V–T *surface in the* p–T *plane. The full lines are phase boundaries*

the *saturation vapour pressure* or *equilibrium vapour pressure* of the substance at the particular temperature. Liquids are almost in-compressible, so that a very large increase in pressure is needed to produce even a small decrease in volume over the part DE of the isotherm. At sufficiently high pressures transformation from the liquid to the solid occurs, as at F.

In the temperature range $T_t < T < T_c$ the locus of points such as B is known as the *saturated vapour line* and that of points such as D is called the *saturated liquid line* [2]. These two lines meet at the *critical point* C. Since C is the limiting position of the horizontal portions of the isotherms, it is a point of inflexion and, therefore, the equations

$$\left(\frac{\partial p}{\partial V}\right)_{T=T_c} = \left(\frac{\partial^2 p}{\partial V^2}\right)_{T=T_c} = 0 \qquad (13.1)$$

define C. Also, the critical point is the state in which the saturated liquid and saturated vapour become identical.

At the temperature T_t, which is less than T_c, the substance is gaseous at low pressure, but as the pressure is increased, a state is reached in which both liquid and solid are produced and solid, liquid and gas coexist in equilibrium. The line HJ in Figure 13.2 is known as the *triple line* [3]. Further compression takes place at constant pressure until all the substance is in the solid phase and then the isotherm rises sharply.

When the temperature is below T_t, condensation from the gaseous phase is directly to the solid phase. This condensation takes place at a constant pressure, which is the equilibrium vapour pressure of the solid. The direct change from solid to gas is known as *sublimation*.

The *p–T* diagram (Figure 13.3) shows the phases of the pure substance. In this diagram two coexistent phases are represented by a line and three coexistent phases by the point of intersection of three lines. In Figure 13.3 the sublimation curve represents the states in which solid and gas coexist in equilibrium, the fusion line those states in which solid and liquid coexist in equilibrium, and the vaporisation curve those in which liquid and gaseous phases coexist in equilibrium. For substances that expand on melting the fusion curve has a positive slope, as shown in Figure 13.3. In contrast, for substances that contract on melting, such as water, the fusion curve has a negative slope.

13.2 Equations of state for real gases

The *p–V–T* surface for a pure substance is sufficiently complex to make its complete representation by a reasonably simple analytical equation of state very unlikely. A more fruitful investigation is likely to be into the form of an analytical equation of state for the gaseous phase, since it is known that, in the limit of vanishingly small pressures, the equation of state then has the simple form $pV_m = RT$, where V_m is the molar volume and R is the gas constant.

One approach to the problem of devising equations of state for real gases is to recall that, in Section 6.1, it was shown that the behaviour of a real gas could be represented over a very wide range of temperature by an equation of the form

$$pV_m = A + Bp + Cp^2 + \cdots \qquad (13.2)$$

where A, B, C, ... are functions of temperature and the nature of the gas. For an explicit equation of state it is necessary to give analytic form to A, B, C, ... This may be done by considering the conditions that must be satisfied by an equation of state for a real gas, as in the example at the end of the chapter. Examples of these conditions are the following:

1 The equation of state should reduce to the ideal gas equation in the limit of vanishingly small pressures, at all temperatures — that is,

$$\lim_{p \to 0} \left(\frac{pV_m}{RT} \right) = 1, \text{ at all temperatures} \qquad (13.3)$$

2 There must be a critical isotherm, corresponding to a temperature T_c, that shows a point of inflexion — that is,

$$\left(\frac{\partial p}{\partial V} \right)_{T = T_c} = \left(\frac{\partial^2 p}{\partial V^2} \right)_{T = T_c} = 0 \qquad (13.4)$$

3 At the Boyle temperature T_B the slope of the graph of pV against p tends to zero in the limit of vanishingly small pressures (see Section 6.1) — that is,

$$\lim_{p \to 0} \left[\frac{\partial(pV)}{\partial p} \right]_{T = T_B} = 0 \qquad (13.5)$$

4 The equation of state should predict a *Joule–Thomson effect* (see Section 16.2) and give a value for the *Joule–Thomson coefficient* μ, where

$$\mu = \frac{1}{C_p} \left[T \left(\frac{\partial V}{\partial T} \right)_p - V \right] \qquad (13.6)$$

Writing

$$\frac{pV_m}{RT} = Z$$

known as the *compressibility factor* of the gas, μ can be written

$$\mu = \frac{RT^2}{pC_p} \left(\frac{\partial Z}{\partial T} \right)_p \qquad (13.7)$$

An alternative approach to the construction of equations of state is to start with the molar internal energy U_m and enthalpy H_m for the system. From the first law of thermodynamics

$$dU_m = TdS_m - p \, dV_m \qquad (13.8)$$

so that

$$\left(\frac{\partial U}{\partial V} \right)_T = T \left(\frac{\partial S}{\partial V} \right)_T - p$$

or, using Maxwell relation (M3) (page 113),

$$\left(\frac{\partial U}{\partial V} \right)_T = T \left(\frac{\partial p}{\partial T} \right)_V - p \qquad (13.9)$$

This is, of course, the energy equation. Enthalpy is defined as $H = U + pV$, so that

$$dH_m = dU_m + pdV_m + V_m dp$$

or, substituting for dU_m from Equation (13.8),

$$dH_m = TdS_m + V_m dp$$

Therefore,

$$\left(\frac{\partial H}{\partial p}\right)_T = T\left(\frac{\partial S}{\partial p}\right)_T + V_m$$

which, using Maxwell relation (M1) (page 114), becomes

$$-\left(\frac{\partial H}{\partial p}\right)_T = T\left(\frac{\partial V}{\partial T}\right)_p - V_m \qquad (13.10)$$

Equations (13.9) and (13.10) are thermodynamic equations of state. They may be combined with Equation (11.20) to give

$$\left[p + \left(\frac{\partial U}{\partial V}\right)_T\right]\left[V_m - \left(\frac{\partial H}{\partial p}\right)_T\right] = T(C_{p,m} - C_{V,m}) \qquad (13.11)$$

When expressions for $U_m(T,V)$ and $H_m(T,p)$ are known, the equation of state may be obtained from Equation (13.11).

Unfortunately, U_m and H_m are not usually known exactly and further manipulation of these equations is needed to obtain approximate equations of state. Equation (13.9) may be written

$$\left(\frac{\partial U}{\partial V}\right)_T = T^2\left[\frac{\partial(p/T)}{\partial T}\right]_V$$

which, on integrating with V constant, gives

$$\frac{p}{T} = \int\left(\frac{\partial U}{\partial V}\right)_T \frac{dT}{T^2} + \Lambda f(V_m) \qquad (13.12)$$

where the function of integration has been written $\Lambda f(V_m)$. Equation (13.12) may be written

$$p - T\int\left(\frac{\partial U}{\partial V}\right)_T \frac{dT}{T^2} = \Lambda T f(V_m)$$

Putting

$$\varphi(V_m) = 1/f(V_m)$$

and

$$\psi(T, V_m) = -T \int \left(\frac{\partial U}{\partial V}\right)_T \frac{dT}{T^2} \quad \text{(integration at constant } V\text{)}$$

gives

$$[p + \psi(T, V_m)] \, \varphi(V_m) = \Lambda T \tag{13.13}$$

Any pressure-explicit equation of state — that is, an equation of the form $p = p(T, V_m)$ — must be expressible in the form of Equation (13.13) if it is to be consistent with the laws of thermodynamics. The equation of state can be obtained from Equation (13.12) when exact expressions for $U_m(T, V_m)$ and $\varphi(V_m)$ are available. It should be noted that $\varphi(V_m)$ and $\psi(T, V_m)$ are not, in fact, independent. As T approaches absolute zero, the third law of thermodynamics (see Chapter 15) indicates that

$$\lim_{T \to 0} \left(\frac{\partial S}{\partial V}\right)_T = \lim_{T \to 0} \left(\frac{\partial p}{\partial T}\right)_V = 0$$

so that

$$\frac{\Lambda}{\varphi(V_m)} = \lim_{T \to 0} \left(\frac{\partial \psi}{\partial T}\right)_V$$

This constraint is only important near absolute zero; at higher temperatures $\varphi(V_m)$ and $\psi(T, V_m)$ may be treated as independent.

13.3 van der Waals' equation

As a first step in producing an equation of state from Equation (13.13), put $\varphi(V_m)$ equal to V_m and take $U_m(T, V_m)$ to be independent of volume — that is, $U_m = U_m^0(T)$. Then $\psi(T, V_m)$ is zero and Equation (13.13) becomes

$$pV_m = RT \tag{13.14}$$

where R has been written for Λ. Equation (13.14) is the equation of state of the ideal gas. As is shown in Chapter 17, Equation (13.14) can be deduced on a kinetic theory model if, among others, the assumptions are made of vanishingly small molecular volume (i.e. very-short-range repulsive inter-molecular forces) and negligible intermolecular attractions. Equations that are more realistic than Equation (13.14) should attempt to allow for the intermolecular forces (both attractive and repulsive) while still satisfying Equation (13.13).

As a first attempt to allow for the finite volume of the molecules, $\varphi(V_m)$ may be written as $\varphi(V_m) = V_m - b$, where, for a particular gas, b is a very slowly varying function of temperature only. The presence of attractive forces between the molecules causes the internal energy to decrease as the volume increases, so that the next stage of approximation is to write

$$U_m = U_m^0(T) - \frac{a}{V_m} \qquad (13.15)$$

where a depends on the temperature and the nature of the gas. Equation (13.15) gives

$$\psi(T,V_m) = a/V_m^2$$

so that Equation (13.13) becomes

$$\left(p + \frac{a}{V_m^2}\right)(V_m - b) = RT \qquad (13.16)$$

This is *van der Waals' equation of state.*

Isotherms given by Equation (13.16) projected in the p–V plane are shown in Figure 13.4. It can be seen that the equation predicts a critical temperature T_c and a critical point C. Also, there is reasonable qualitative agreement with Figure 13.2 for those states where the substance is either wholly liquid or wholly gas. However, for states where two phases coexist in equilibrium, van der Waals isotherms indicate that an increase in pressure produces an increase in volume. Also, at sufficiently low temperatures, the pressure becomes negative. This last feature has been interpreted as denoting the liquid phase under tension. The maximum value of this tension, for a given temperature, is then interpreted as the tensile strength of the liquid phase at that temperature.

Since the critical isotherm passes through a point of inflexion at the critical point, Equations (13.4), must hold at C. When applied to a gas obeying van der Waals' equation, these equations give

$$\left(\frac{\partial p}{\partial V}\right)_T = - \frac{RT}{(V_m-b)^2} + \frac{2a}{V_m^3}$$

and

$$\left(\frac{\partial^2 p}{\partial V^2}\right)_T = \frac{2\,RT}{(V_m-b)^3} - \frac{6a}{V_m^4}$$

The conditions for the critical point are then

$$\frac{RT_c}{(V_{c,m}-b)^2} = \frac{2a}{V_{c,m}^3} \qquad (13.17)$$

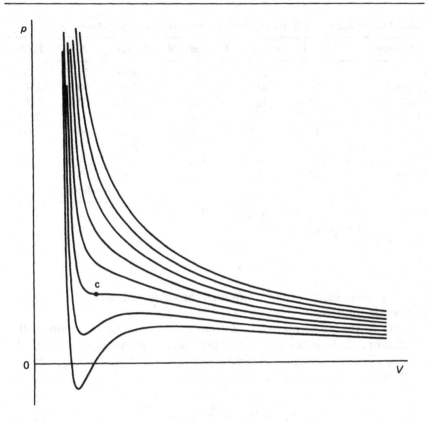

Figure 13.4 *The isotherms for a van der Waals gas. C is the critical point*

and

$$\frac{2\,RT_c}{(V_{c,m}-b)^3} = \frac{6a}{V_{c,m}^4} \tag{13.18}$$

where $V_{c,m}$ is the molar volume at the critical point. Solving Equations (13.16)–(13.18) gives

$$V_{c,m} = 3b; \quad T_c = \frac{8a}{27Rb}; \quad p_c = \frac{a}{27b^2} \tag{13.19}$$

It can readily be seen from Equations (13.19) that the quantity $RT_c/p_cV_{c,m}$ has no dimensions and does not contain either of the parameters a and b. Therefore, its numerical value of 8/3, or 2.67, should be the same for all gases obeying van der Waals' equation. Some experimental values of $RT_c/p_cV_{c,m}$ are given in Table 13.1, from which it can be seen that the value

Table 13.1 Characteristic parameters of some common substances

Substance	T_c/K	p_c/10^5Pa	$V_{c,m}$/cm^{3}mol^{-1}	$RT_c/p_cV_{c,m}$	T_t/K	T_B/T_c
Neon	44.4	27.3	41.7	3.25	24.57	2.74
Argon	150.7	48.6	75.2	3.43	83.78	2.72
Hydrogen	32.99	12.9	65.5	3.24	13.84	3.46
Oxygen	154.8	50.8	78	3.25	54.3	2.62
Nitrogen	126.2	33.9	90.1	3.44	63.1	2.60
Carbon dioxide	304.2	73.8	94.0	3.65	216.55	
Sulphur dioxide	430.6	78.8	122	3.72	197.6	
Water	647.3	221.2	59.1	4.12	273.16 (definition)	

Values of T_c, p_c, $V_{c,m}$ from Kaye and Laby (1973).
Values of T_B from Roberts and Miller (1951).
Values of T_t from Hsieh (1975).

does vary from gas to gas and, although the value is of the correct order of magnitude for a van der Waals gas, it is no more.

The Boyle temperature T_B of a gas is the temperature at which the coefficient B in Equation (13.2) has the value zero. An expression for T_B may be obtained in terms of a and b for a van der Waals gas as follows.

Van der Waals' equation may be written

$$pV_m = RT - \frac{a}{V_m} + bp + \frac{ab}{V_m^2}$$

At moderate values of pressure and volume, V_m is given approximately by the ideal gas equation — that is,

$$V_m = RT/p$$

Substituting this value of V_m in van der Waals' equation gives, approximately [4],

$$pV_m = RT + \left(b - \frac{a}{RT}\right)p + \frac{ab\, p^2}{R^2T^2} \tag{13.20}$$

Comparing Equation (13.20) with Equation (13.2) gives $B = (b - a/RT)$. Therefore, for a van der Waals gas,

$$T_B = \frac{a}{Rb} \tag{13.21}$$

The value of T_c for a van der Waals gas is $8a/27Rb$, so that, for such a gas,

$$\frac{T_B}{T_c} = \frac{27}{8} = 3.38 \tag{13.22}$$

Table 13.1 shows values of T_B/T_c for some real gases; it can be seen that the values are not very close to that given by Equation (13.22).

The results given in Table 13.1 show clearly that van der Waals' equation is no more than a reasonable approximation to the equation of state of real gases. Better equations of state are needed and these require a better knowledge of the dependence of $\varphi(V_m)$ on V_m and of U_m on T and V_m. Experiment suggests that both $\varphi(V_m)$ and U_m can be written as a series, the respective forms being

$$\varphi(V_m) = V_m + \cdots$$

and

$$U_m = U_m^0(T) + \frac{A(T)}{V_m} + \cdots \tag{13.23}$$

and much effort has gone into extending these expansions.

An example of a widely used improved equation of state that satisfies Equation (13.13) is the *Beattie–Bridgeman equation*:

$$p = \frac{RT}{V_m}\left(1 - \frac{c}{V_m T^3}\right)\left[1 + \frac{B_0}{V_m}\left(1 - \frac{b}{V_m}\right)\right] - \frac{A_0(1 - a/V_m)}{V_m^2} \tag{13.24}$$

This equation has five constants, A_0, B_0, a, b, c, which must be determined by experiment.

13.4 Reduced equations of state

It is sometimes helpful to express equations of state in terms of what are called *reduced coordinates* — that is, the coordinates expressed, using their values at the critical point as units. Then the reduced temperature $\hat{T}$, reduced pressure $\hat{p}$ and reduced molar volume $\hat{V}_m$ are given, respectively, by

$$\hat{T} = \frac{T}{T_c}; \ \hat{p} = \frac{p}{p_c}; \ \hat{V}_m = \frac{V_m}{V_{c,m}} \tag{13.25}$$

Using reduced coordinates, van der Waals' equation becomes

$$\left(\hat{p}\,p_c + \frac{a}{\hat{V}_m^2 V_{c,m}^2}\right)(\hat{V}_m V_{c,m} - b) = R\hat{T}T_c \tag{13.26}$$

Substituting for the critical coordinates in terms of a, b and R from Equations (13.19) gives

$$\left(\frac{\hat{p}\,a}{27b^2} + \frac{a}{\hat{V}_m^2\,9\,b^2}\right)(3b\hat{V}_m - b) = \frac{R\hat{T}8a}{27Rb}$$

which simplifies to

$$\left(\hat{p} + \frac{3}{\hat{V}_m^2}\right)(3\hat{V}_m - 1) = 8\hat{T} \qquad (13.27)$$

Equation (13.27) is known as the *reduced equation of van der Waals*. One obvious deduction from this equation is that, for all van der Waals gases, an isotherm corresponding to a given value of $\hat{T}$ should have the same form when plotted in the $\hat{p}$–$\hat{V}_m$ plane. This result is found to hold approximately for real gases and is known as the *law of corresponding states*.

The reduced equation offers a convenient way of obtaining the locus of the Boyle points for a van der Waals gas. This locus is obtained from the condition

$$\left(\frac{\partial \hat{p}\hat{V}}{\partial \hat{p}}\right)_{\hat{T}} = 0 \qquad (13.28)$$

To obtain the required locus, Equation (13.27) must be written in a form that relates $\hat{p}\hat{V}$ to $\hat{p}$. This may be done by writing the equation in the form

$$(\hat{p}\hat{V}_m^2 + 3)(3\hat{V}_m - 1) = 8\hat{T}\hat{V}_m^2$$

and then multiplying both sides by $\hat{p}^2$ to give

$$(\hat{p}\,\hat{p}^2\,\hat{V}_m^2 + 3\hat{p}^2)(3\hat{V}_m - 1) = 8\hat{T}\,\hat{p}^2\hat{V}_m^2 \qquad (13.29)$$

Put $y = \hat{p}\hat{V}_m$ and $x = \hat{p}$ in the above equation. Then

$$(xy^2 + 3x^2)(3y/x - 1) = 8\hat{T}y^2$$

and, therefore,

$$\left(\frac{\partial y}{\partial x}\right)_{\hat{T}} = \frac{y^2 - 9y + 6x}{9y^2 - 2xy - 16\hat{T}y + 9x} \qquad (13.30)$$

Applying condition (13.28) to Equation (13.30) shows that the locus of the Boyle points is given by

$$y^2 - 9y + 6x = 0 \qquad (13.31)$$

which is a parabola.

The Boyle temperature T_B is that for which $(\partial y/\partial x)_{\hat{T}}$ is zero in the limit as x tends to zero. From Equation (13.31), therefore, $\hat{T}_B$ for a van der

Waals gas is given by $y = 9$ and $x = 0$. Equation (13.27) gives the corresponding value of $\hat{T}$ as 27/8 — that is, T_B for a van der Waals gas is $27T_c/8$. Since T_c is $8a/27Rb$, T_B is a/Rb, which agrees with Equation (13.21).

■ EXAMPLE

Q. A certain closed hydrostatic system has as its equation of state

$$\frac{T}{p} = a_0 + a_1 V_m + a_2\frac{V_m^2}{T} + a_3\frac{V_m^3}{T}$$

where V_m is the molar volume when the temperature is T and the pressure is p, and a_0, a_1, a_2 and a_3 are constants. Deduce as much as possible about the constants if the following conditions are satisfied.

(a) The substance behaves as an ideal gas at high temperatures.
(b) At low temperatures the limiting molar volume is b.
(c) The critical temperature is T_c.

A. At sufficiently high temperatures both the terms $a_2 V_m^2/T$ and $a_3 V_m^3/T$ become negligible and the equation of state takes the form

$$T = pa_0 + pa_1 V_m \qquad (13.32)$$

Comparing this equation with the ideal gas equation,

$$T = \frac{pV_m}{R}$$

gives

$$a_0 = 0 \text{ and } a_1 = \frac{1}{R} \qquad (13.33)$$

so that the equation of state may be written

$$\frac{T}{p} = \frac{V_m}{R} + a_2\frac{V_m^2}{T} + a_3\frac{V_m^3}{T} \qquad (13.34)$$

Putting the above equation in the form

$$\frac{T^2}{p} = \frac{TV_m}{R} + a_2 V_m^2 + a_3 V_m^3 \qquad (13.35)$$

it can be seen that, in the limit as T tends to zero,

$$a_2 V_m^2 = - a_3 V_m^3$$

The limiting value of V_m at low temperatures is b, so that

$$a_2 = -a_3 b \qquad (13.36)$$

The existence of a critical temperature T_c implies the following conditions:

$$\left(\frac{\partial p}{\partial V}\right)_{T = T_c} = \left(\frac{\partial^2 p}{\partial V^2}\right)_{T=T_c} = 0 \qquad (13.37)$$

The appropriate partial derivatives from Equation (13.34) are

$$-\frac{T}{p^2}\left(\frac{\partial p}{\partial V}\right)_T = \frac{1}{R} + \frac{2a_2 V_m}{T} + \frac{3a_3 V_m^2}{T}$$

and

$$\frac{2T}{p^3}\left(\frac{\partial^2 p}{\partial V^2}\right)_T = \frac{2a_2}{T} + \frac{6a_3 V_m}{T}$$

At the critical point $T = T_c$ and $V_m = V_{c,m}$, so that

$$\frac{1}{R} + \frac{2a_2 V_{c,m}}{T_c} + \frac{3a_3 V_{c,m}^2}{T_c} = 0 \qquad (13.38)$$

and

$$\frac{2a_2}{T_c} + \frac{6a_3 V_{c,m}}{T_c} = 0 \qquad (13.39)$$

Substituting $a_2 = - a_3 b$ into Equation (13.39) gives

$$V_{c,m} = \frac{b}{3} \qquad (13.40)$$

and using this value in Equation (13.38), together with $a_2 = - a_3 b$, gives

$$\frac{1}{R} - \frac{2a_3 b^2}{3T_c} + \frac{3a_3 b^2}{9T_c} = 0$$

or

$$a_3 = \frac{3T_c}{b^2 R}$$

Therefore,

$$a_2 = -\frac{3T_c}{bR}$$

and the final form of the equation of state is

$$\frac{T}{p} = \frac{V_m}{R} - \frac{3T_cV_m^2}{bR\,T} + \frac{3\,T_cV_m^3}{b^2R\,T}$$

▮ EXERCISES

1 A closed hydrostatic system occupies a molar volume V_m when the pressure is p and the temperature T. Test the following equations of state of such a system for thermodynamic validity.

(a) *Berthelot's equation*

$$\left(p + \frac{a'}{TV_m^2}\right)(V_m - b') = RT$$

R is the gas constant and a' and b' are constants.

(b) The *Redlich–Kwong equation*

$$\left(p + \frac{a''}{V_m(V_m + b'')\sqrt{T}}\right)(V_m - b'') = RT$$

where a'' and b'' are constants.

(c) *Dieterici's equation*

$$p(V_m - b''') = RT \exp\left(-a'''/RTV_m\right)$$

where a''' and b''' are constants.

2 Obtain expressions for the critical coordinates of a Berthelot gas (see Exercise 1(a)) and derive the reduced equation of state.

3 Show that if the equation for a closed hydrostatic system

$$pV_m = A + Bp + Cp^2 + \cdots$$

is written in the form

$$pV_m = A + \frac{B'}{V_m} + \frac{C'}{V_m^2} + \cdots$$

the coefficients are related approximately by the equations

$$B = \frac{B'}{A}$$

$$C = \frac{C'A - B'^2}{A^3}$$

4   Derive an expression for the difference between the principal molar heat capacities of a van der Waals gas.

5   Deduce a value for the Boyle temperature of a Berthelot gas in terms of its critical temperature.

6   Determine the values of a'' and b'' in the Redlich–Kwong equation in terms of p_c and T_c.

7   The heat capacity of a liquid on the liquid saturation curve C_{sat} is the heat capacity of a liquid that is maintained at all temperatures in equilibrium with an infinitesimal amount of vapour. If the system occupies a volume V when the temperature is T and the pressure is p, show that

$$C_{sat} = C_V + T\left(\frac{\partial p}{\partial T}\right)_V \left(\frac{\partial V}{\partial T}\right)_{sat}$$

where C_V is the principal heat capacity at constant volume.

8   For an equilibrium state of a mixture of liquid and vapour, such as that represented by the point K in Figure 13.2, show that the mass of liquid m_l is related to the mass of vapour m_g by

$$\frac{m_l}{m_g} = \frac{KB}{KD}$$

Hint: Remember that the total mass of the system is constant, wherever K is located on BD, and that the density of each phase remains constant along BD.

Notes

1   The terms 'gas' and 'liquid' have been used here in a classical sense: liquids have a definite volume at a given pressure and temperature; gases fill completely the volume available to them at all temperatures. Strictly, however, the terms 'liquid' and 'gas' are needed only when the phases coexist; then the less dense phase is the gas and the denser is the liquid. However, it is possible for the substance to pass from a state where it is obviously 'liquid' to one where it is 'gas' without condensing — for example, by means of the dotted path linking L and G in Figure

13.3. This indicates that gases and liquids are both fluids that differ only in degree. However, in this book the term 'liquid' will be used for the condition of a pure substance represented by points that lie between the fusion curve and the vaporisation curve in the temperature range T_t–T_c.

2   A saturated liquid is a liquid in equilibrium with an infinitesimal amount of vapour; a saturated vapour is a vapour in equilibrium with an infinitesimal amount of liquid.

3   Helium is exceptional in showing a triple surface rather than a line.

4   This approach is also useful when it is required to obtain V_m from van der Waals' equation, given p and T.

Use the ideal gas equation to find an initial approximate value of V_m. Then write van der Waals' equation in the form

$$\frac{pV_m}{RT} = 1 + \frac{pb}{RT} - \frac{a}{RTV_m} + \frac{ab}{RTV_m^2}$$

and use the approximate value of V_m in the right-hand side of the above equation to obtain a new value for pV_m/RT and, therefore, for V_m. A few iterations are needed to obtain a satisfactory answer, but this is usually quicker than solving the general equation for a cubic.

14

Phase changes

Each equilibrium state of a closed hydrostatic system is characterised by a set of values of pressure p, volume V and temperature T. When the system consists of a single chemical species (a *one-component system*), for some ranges of temperature and pressure only a single phase [1] is present — that is, solid, liquid or gas. However, for other ranges of pressure and temperature two phases coexist in equilibrium, while along the triple line (see Section 13.1) all three phases coexist in equilibrium.

Examination of the p–V–T surface for a one-component system (Figure 13.1) shows that the processes of melting and vaporisation occur under conditions of constant temperature when the pressure is held constant, but the specific volumes of the respective phases are different. This chapter discusses some thermodynamic aspects of such 'ordinary' phase changes.

14.1 Equilibrium between phases of a closed hydrostatic system

Consider a one-component closed hydrostatic system in a state in which two phases coexist in equilibrium. If surface effects across the phase boundary are neglected, both pressure and temperature are uniform throughout both phases. In the p–T diagram (Figure 13.3) this equilibrium state is represented by a point on the vaporisation curve when the two phases are liquid and vapour, by a point on the fusion curve when the phases are liquid and solid, and by a point on the sublimation curve when the phases are solid and vapour.

In a general situation where two phases coexist in equilibrium, let the more dense phase be designated phase 1 and the less dense phase 2. The equilibrium state is represented by a point on the appropriate phase boundary in the p–T diagram, as in Figure 14.1. Let an infinitesimal mass of the substance change from phase 1 to phase 2 by a reversible process, which must take place at constant temperature and pressure. Since the change dG in the Gibbs function of the complete system is given by

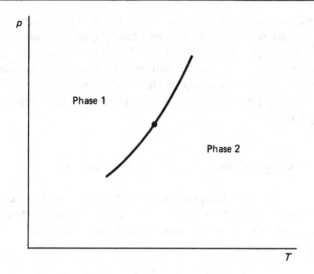

Figure 14.1 *Part of a phase boundary plotted in the p–T plane*

$$dG = Vdp - SdT \qquad (14.1)$$

for any process (see Section 10.6), there can be no change in G during a phase change with p and T constant.

Neglect interface effects and let m_1 and m_2 be, respectively, the masses of substance in phases 1 and 2. Then

$$G = G_1 + G_2$$

where $G_1 = m_1 g_1$ and $G_2 = m_2 g_2$, and g_1 and g_2 are the specific Gibbs functions of phases 1 and 2, respectively. The expression for G may then be written

$$G = m_1 g_1 + m_2 g_2 \qquad (14.2)$$

Now g may be written as a function of temperature and pressure only, since it is a function of state. Therefore, if m_1 increases by dm_1 and m_2 by dm_2 in an infinitesimal phase change taking place at constant temperature and pressure,

$$dG = g_1 dm_1 + g_2 dm_2 = 0 \qquad (14.3)$$

Since the system is closed, the total mass must remain constant and, therefore,

$$dm_1 + dm_2 = 0 \qquad (14.4)$$

Combining Equations (14.3) and (14.4) gives

$$g_1 = g_2 \qquad (14.5)$$

which is the condition for two phases to coexist in equilibrium — that is, the specific Gibbs functions of the two phases must be equal. Since the value of G does not vary as m_1, say, changes, the mass of phase 1, coexisting with phase 2 under conditions of constant temperature and pressure, can vary from zero up to the total mass of the system. This explains the appearance of the isotherms in the two-phase regions in the p–V–T diagram.

14.2 The Clapeyron–Clausius equation

Consider now how the temperature at which two phases coexist in equilibrium depends on the pressure. When the system is in the equilibrium state, represented by the point a in Figure 14.2, the equilibrium condition is

$$g_1(a) = g_2(a) \qquad (14.6)$$

If the pressure is now increased by dp and the two phases continue to coexist in equilibrium, the temperature of the system must increase by dT to ensure that the new equilibrium state, represented by the point b in Figure 14.2, remains on the appropriate phase line. The new equilibrium condition is

$$g_1(b) = g_2(b)$$

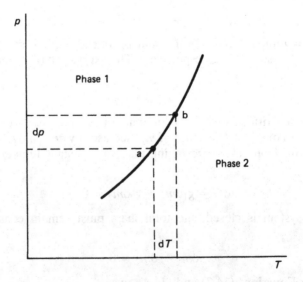

Figure 14.2 *The effect of change of pressure on the temperature for two phases in equilibrium*

so that

$$dg_1 = dg_2 \qquad (14.7)$$

Since g may be treated as a function of temperature and pressure (i.e. $g = g(T, p)$), Equation (14.7) may be written

$$dg_1 = \left(\frac{\partial g_1}{\partial T}\right)_p dT + \left(\frac{\partial g_1}{\partial p}\right)_T dp$$

$$= \left(\frac{\partial g_2}{\partial T}\right)_p dT + \left(\frac{\partial g_2}{\partial p}\right)_T dp = dg_2 \qquad (14.8)$$

Now, by comparison with Equation (14.1), dg may be written

$$dg = v dp - s dT \qquad (14.9)$$

where s is the specific entropy and v is the specific volume. Then

$$\left(\frac{\partial g}{\partial T}\right)_p = -s \text{ and } \left(\frac{\partial g}{\partial p}\right)_T = v \qquad (14.10)$$

Substituting from Equation (14.10) into Equation (14.8), using the appropriate suffixes, gives

$$(s_2 - s_1) \, dT = (v_2 - v_1) \, dp$$

or

$$\frac{dp}{dT} = \frac{s_2 - s_1}{v_2 - v_1} \qquad (14.11)$$

Now $g = u + pv - Ts$ and, in equilibrium, $g_1 = g_2$. Therefore,

$$s_2 - s_1 = \frac{(u_2 + pv_2) - (u_1 + pv_1)}{T} = \frac{h_2 - h_1}{T}$$

where h is the specific enthalpy. Substituting for $s_2 - s_1$ in Equation (14.11) gives

$$\frac{dp}{dT} = \frac{h_2 - h_1}{T(v_2 - v_1)} = \frac{h_{12}}{T(v_2 - v_1)} \qquad (14.12)$$

h_{12} is the *specific enthalpy of the phase change*. It represents the energy that must be supplied at constant temperature and pressure to convert unit mass of phase 1 into phase 2 and is commonly called the *specific latent heat* [2] of the phase change.

Equation (14.12) was deduced by Clapeyron but first proved rigorously by Clausius. It is known as the Clapeyron–Clausius equation or, sometimes, simply the equation of Clapeyron. It is widely used to predict the effect of change of pressure on phase transition temperatures and also to calculate h_{12} for phase transitions.

The Clapeyron–Clausius equation is only able to describe phase changes in which there is a specific enthalpy ($h_{12} > 0$) accompanied by a change in specific volume ($v_1 \neq v_2$). Since h_{12} is given by $T(s_2 - s_1)$, it follows from Equations (14.10) that the phase changes described by the Clapeyron–Clausius equation are characterised by discontinuous changes in the first derivatives of the specific Gibbs function g — that is,

$$\left(\frac{\partial g}{\partial T}\right)_p \text{ and } \left(\frac{\partial g}{\partial p}\right)_T$$

both change discontinuously in going from phase 1 to phase 2. Following Ehrenfest, phase changes may be classified by the order of the lowest derivative of g which shows a discontinuity at the transition. The transitions of melting and vaporisation are classified as *first-order phase changes*.

Equation (14.12) may also be used to determine values of thermodynamic temperature T when measurements of h_{12} are made using an empirical temperature scale θ. Assuming a monotonic relation between T and θ, Equation (14.12) may be written

$$h_{12} = T(v_2 - v_1)\frac{dp}{d\theta} \cdot \frac{d\theta}{dT}$$

which gives

$$\ln T = \int \frac{v_2 - v_1}{h_{12}} \frac{dp}{d\theta} \, d\theta + \text{constant} \tag{14.13}$$

The value of the constant may be determined from the value of temperature assigned to the triple-point temperature of water and, therefore, T may be determined as a function of θ.

14.3 The equation of the vaporisation curve

Equation (14.12) gives the differential form of the appropriate phase boundary. To obtain an explicit equation for the phase boundary, Equation (14.12) must be integrated. This may be done approximately for the vaporisation curve and sublimation curve using some simple assumptions.

Consider the transition from liquid to vapour. Assume that the vapour obeys the ideal gas equation and that the molar volume of the vapour $V_{m,g}$ is very much larger than that of the liquid $V_{m,l}$. Then, for 1 mol of each phase, these assumptions may be written

$$pV_{m,g} = RT$$

and

$$V_{m,g} \gg V_{m,l}$$

Letting phase 2 be the vapour phase and phase 1 be the liquid phase, Equation (14.12) becomes

$$\frac{dp}{dT} \approx \frac{H_{m,v}}{TV_{m,g}} = \frac{H_{m,v}p}{T^2R} \tag{14.14}$$

where $H_{m,v}$ is the molar enthalpy, or molar latent heat, of vaporisation. This equation is often used to estimate $H_{m,v}$ at low pressures.

To integrate Equation (14.14), the dependence of $H_{m,v}$ on T needs to be known. The simplest assumption is to treat $H_{m,v}$ as a constant. Then

$$\ln p = -\frac{H_{m,v}}{RT} + \text{constant}$$

or

$$p = \text{constant} \cdot \exp\left(-H_{m,v}/RT\right) \tag{14.15}$$

Experimental results for the equilibrium between liquid and gaseous water are shown in Figure 14.3, from which it can be seen that Equation (14.15) is a reasonable representation of the behaviour over the temperature range 270–430 K.

Equation (14.14) may also be applied to the sublimation curve. Assuming that the molar enthalpy of sublimation $H_{m,s}$ is a constant, the equation for the sublimation curve is then

$$p = \text{constant} \cdot \exp\left(-H_{m,s}/RT\right) \tag{14.16}$$

A closer examination of the measured values of $H_{m,v}$ suggests that, except for temperatures close to T_c, a better approximation is to treat $H_{m,v}$ as varying uniformly with temperature — that is, to write

$$H_{m,v} = H_{m0} + nT$$

where H_{m0} and n are constants. Equation (14.14) then gives

$$\frac{dp}{dT} = \frac{(H_{m0} + nT)p}{T^2R}$$

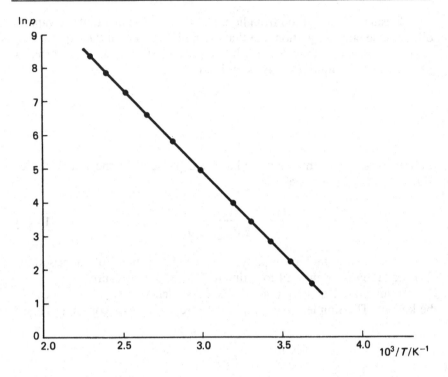

Figure 14.3 *The saturation vapour pressure* p *of water as a function of temperature* T

which integrates to give

$$\ln p = -\frac{H_{m0}}{R} + \frac{n \ln T}{R} + \text{constant} \qquad (14.17)$$

This has the form of Equation (14.15), but with a correction term.

14.4 The Clausius equation

The discussion of Section 14.2 shows the importance of knowing dh_{12}/dT if the Clapeyron–Clausius equation is to be integrated, and this suggests that a direct investigation of this dependence is needed.

In any first-order phase change let a mass m change from phase 1 to phase 2 under the appropriate constant temperature and pressure. The specific enthalpy change for the process is h_{12}, where

$$h_{12} = (h_2 - h_1) = T(s_2 - s_1)$$

Here s is the specific entropy. When the pressure of the system changes, the temperature must change to keep the state of the system on the appropriate phase boundary. This removes one of the degrees of freedom of the system, which means that, subject to this condition, both h and s are functions of temperature only. The dependence of h on T must then satisfy the equation

$$\frac{d}{dT}\left(\frac{h_2}{T}\right) - \frac{d}{dT}\left(\frac{h_1}{T}\right) = \frac{ds_2}{dT} - \frac{ds_1}{dT} \tag{14.18}$$

or

$$\frac{1}{T}\frac{dh_2}{dT} - \frac{h_2}{T^2} - \frac{1}{T}\frac{dh_1}{dT} + \frac{h_1}{T^2} = \frac{ds_2}{dT} - \frac{ds_1}{dT}$$

Therefore,

$$\frac{d(h_2 - h_1)}{dT} - \frac{(h_2 - h_1)}{T} = T\frac{ds_2}{dT} - T\frac{ds_1}{dT}$$

or

$$\frac{dh_{12}}{dT} - \frac{h_{12}}{T} = c_2 - c_1 \tag{14.19}$$

This is the equation of Clausius. c_1 and c_2 may be interpreted as specific heat capacities, subject to the constraint that the two phases remain in equilibrium with each other at constant temperature and pressure. For example, if the phase change considered is from liquid to vapour, c_2 is the *specific heat capacity of the saturated vapour* and c_1 that of the *saturated liquid*.

The condition that c_2 is the specific heat capacity of a saturated vapour — that is, of a vapour in equilibrium with an infinitesimal amount of liquid, leads to some interesting possibilities. Heat capacities are normally measured by supplying a quantity of heat to a system, under appropriate conditions, and measuring the temperature rise produced. When the temperature of a saturated vapour is raised, the volume remaining constant, the vapour ceases to be saturated. To maintain the vapour in a saturated condition, the volume must be decreased as the temperature is raised. Compression involves the performance of work on the system and this work also raises the temperature. The amount of energy transferred as work may be less than the total amount needed to change the temperature of the vapour while keeping it saturated; the value of c_2 in Equation (14.19) is then positive. When the amount of work done on the system to keep the vapour saturated is exactly that needed to raise the temperature,

c_2 is zero. However, the energy transferred as work may be greater than that needed to change the temperature of the vapour while keeping it saturated. Energy must then be extracted as heat from the vapour to keep it saturated and c_2 is then negative. Water vapour comes into this last category.

The left-hand side of Equation (14.19) may be written in the form $Td(h_{12}/T)/dT$, but, even then, the equation cannot be integrated to give the temperature dependence of h_{12} without knowing more about c_1 and c_2. If the assumption is made that c_1 and c_2 are both constant, the result obtained is that

$$h_{12} = (c_2 - c_1)T \ln T + bT$$

where b is a constant.

14.5 The triple point

As can be seen from Figure 13.3, the triple point of a pure substance is determined by the common intersection of the vaporisation curve, the fusion curve and the sublimation curve. From the discussion of Section 14.1 it follows that, at the triple point, the specific Gibbs functions of the solid, liquid and gaseous phases must be equal — that is,

$$g_s = g_l = g_g \qquad (14.20)$$

For a closed hydrostatic system, which has only two degrees of freedom, conditions (14.20) specify a unique thermodynamic state, characterised by a pressure p_t and a temperature T_t, while the coexisting phases have definite densities. It is for this reason that the value of T_t for water was chosen as the standard fixed temperature, rather than the melting point, which was formerly used: Equation (14.13) indicates that the melting point depends on the value of the pressure.

14.6 The critical state

The critical state of a pure substance is represented on the p–V–T surface (Figure 13.1) by that point C at which the density of the liquid phase equals that of the gaseous phase. It was seen in Section 13.1 that C is also the point of inflexion on the critical isotherm, so that, in addition, at the critical point both $(\partial p/\partial V)_T$ and $(\partial^2 p/\partial V^2)_T$ are zero.

These conditions have some interesting consequences. Consider, for example, the behaviour of the specific heat capacity at constant pressure c_p for the substance in the neighbourhood of the critical point. c_p is defined by

$$c_p = T\left(\frac{\partial s}{\partial T}\right)_p$$

where s is the specific entropy of the substance at a temperature T. Using a reciprocal relation, this may be written

$$c_p = T\left(\frac{\partial s}{\partial v}\right)_p\left(\frac{\partial v}{\partial T}\right)_p$$

which, using a reciprocity relation, becomes

$$c_p = - T\left(\frac{\partial s}{\partial v}\right)_p\frac{(\partial p/\partial T)_v}{(\partial p/\partial v)_T}$$

At the critical point $(\partial p/\partial v)_T$ becomes zero, and so c_p should become infinite. Experimentally it is found that c_p has a pronounced maximum as T_c is approached, both from above and below, but for pure substances this is not an infinity. The major experimental problem is that the large value of c_p makes thermal equilibrium difficult to achieve in the neighbourhood of the critical point.

A similar analysis suggests that both the isothermal compressibility κ_T and the cubic expansivity β of the material should become infinite at the critical point. One consequence of a very high value of κ_T is that the gravitational field is able to produce a large density gradient in the vertical direction in a vessel a few centimetres high, giving a non-uniform state of the system. If β is very large, small local changes of temperature produce large fluctuations in the density. In a transparent material these fluctuations give rise to a large amount of light scattering, so that the material has a translucent appearance known as *critical opalescence*. The material may even become completely opaque at the critical point.

These effects make it difficult to maintain a system at the critical point so that precise equilibrium measurements of the critical constants T_c, p_c, $V_{c,m}$ are hard to achieve. The value of T_c may be obtained from the vanishing of the meniscus in a system that has a total density closely equal to the critical density and is contained in a tube. As the temperature is raised from below T_c, the meniscus neither rises nor falls in the container, and becomes less and less distinct, disappearing when T equals T_c. This observation is made difficult by the accompanying critical opalescence, the appearance of which is sometimes used to detect the critical temperature. If the bottom of the tube is connected to a mercury reservoir, the density of the system may be varied and the value of p_c determined at the same time as T_c. The measurement of $V_{c,m}$ is more difficult and is usually achieved by extrapolating the mean of the observed saturated vapour and saturated liquid densities up to T_c. This mean locus, known as the *rectilinear diameter,* is practically linear.

14.7 The determination of the enthalpy of a phase change

A direct determination of the enthalpy of fusion may be made by placing a solid sample of the substance of interest in a calorimeter that also contains a resistor. Energy is supplied to the calorimeter when a current flows in the resistor, and this causes the temperature T to rise until it reaches T_m, the melting point appropriate to the external pressure. The temperature remains at T_m until the whole of the solid is converted to liquid, and then starts to rise again. A graph of T against time t has the form shown in Figure 14.4.

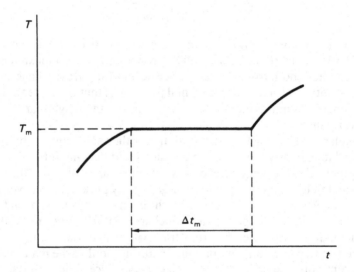

Figure 14.4 *Determination of the enthalpy of fusion from a temperature* T *versus time* t *graph*

If the temperature of the sample is T_m for a time Δt_m, while a current I flows in the resistor under a potential difference E,

$$EI\Delta t_m = mh_f + Q' \qquad (14.21)$$

where m is the mass of the sample, h_f the specific enthalpy of fusion and Q' the heat lost to the surroundings in the interval Δt_m. Q' must be made as small as possible, and can be virtually eliminated by surrounding the calorimeter with a jacket fitted with a separate resistor and maintaining the temperature of this jacket at T_m.

The melting point T_m is the temperature at which the solid and liquid phases are in equilibrium at a given pressure; the locus of melting points is the fusion curve in the graph of pressure against temperature (Figure 13.3).

Melting points of simple materials are not strongly pressure-dependent. The values usually quoted for melting points are measured under a pressure of 1 standard atmosphere when solid, liquid, vapour and external atmosphere are present, and conditions are not strictly those of thermodynamic equilibrium.

The specific enthalpy of vaporisation may be measured directly, using the apparatus shown schematically in Figure 14.5. A sample of the liquid to be studied is placed in a strong container with a curved top, and fitted with a resistor R and a thermometer Th. When a current flows through R, the

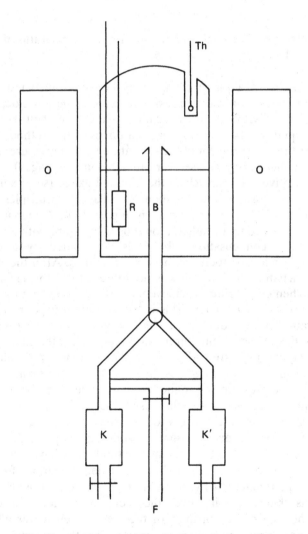

Figure 14.5 *Henning's method for the determination of the enthalpy of vaporisation*

temperature of the liquid rises until boiling starts and the liquid then boils at a steady temperature T_b, determined by the pressure in the apparatus [3]. This pressure can be varied by connecting a pump or a compressor to the tube F. The vapour passes down the tube B, which is fitted with a cone-shaped top to prevent liquid being splashed down. Before conditions are steady the vapour is condensed in the vessel K, but when conditions are steady, condensation is allowed to take place into K', where a mass m condenses in a time Δt_b. An independently controlled oil-bath O at a temperature T_b makes heat exchange with the surroundings very small. Then, if a current I flows through R under a potential difference E,

$$EI\Delta t_b = mh_v \qquad (14.22)$$

where h_v is the specific enthalpy of vaporisation. An apparatus of this type was used by Henning to measure h_v for steam over a wide range of temperature.

The enthalpy of a phase change may be determined indirectly by measuring the saturation vapour pressure and its temperature dependence and using the Clapeyron–Clausius equation, in the form of either Equation (14.14) or Equation (14.15). For liquids a direct static method is usually adequate for the measurement of the saturation vapour pressure. A simple mercury manometer may be used for pressures in the range $0.1 - 10^5$ Pa and more sensitive gauges, such as the McLeod gauge (see Exercise 5 in Chapter 6), for pressures down to about 10^{-3} Pa. The principles involved are illustrated in the apparatus shown in Figure 14.6. A mercury manometer M is connected to a condensation tube C by a tube with an enlarged section D. The condensation tube C is surrounded by a constant-temperature bath at a temperature between T_t and T_c. After the apparatus has been evacuated, sufficient gas is admitted into C for some of it to condense. When equilibrium is achieved, the saturation vapour pressure is measured with the manometer. This is the saturation vapour pressure at the temperature of the liquid in C. The reservoir is then adjusted so that the mercury fills D and the measurement is repeated. If the gas is pure, the saturation vapour pressure indicated by the manometer should be unchanged. For high vapour pressures the open end of the manometer may be closed and the pressure determined from the volume of the trapped air. An alternative procedure uses the result that a liquid boils when its saturation vapour pressure is equal to the external pressure.

The saturation vapour pressure of solids is very small, except at temperatures close to the triple point. Methods that have been used to measure this quantity for metals are based on a determination of the rate of loss of mass of a filament in a vacuum or on the concentration of vapour in an inert gas that is passing over the metal. Langmuir, for example, measured the rate of loss of mass of the metal when it was allowed to evaporate in an evacuated region and assumed that the rate of evaporation was equal to the rate at which vapour molecules would strike the surface in

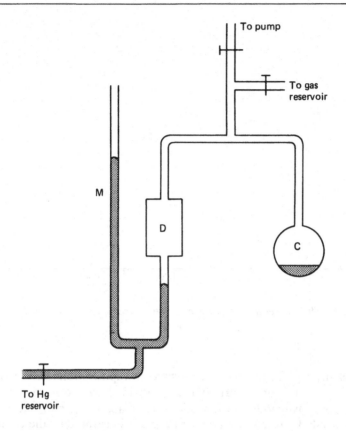

Figure 14.6 *Apparatus to measure the saturation vapour pressure of a liquid*

conditions of thermodynamic equilibrium. The corresponding pressure may be calculated using the kinetic theory of gases (see Section 17.2).

14.8 Results

With regard to the temperature dependence of both the saturation vapour pressure p and the specific enthalpy of vaporisation h_v, the behaviour of water is typical of that of many simple substances. It has already been seen (Figure 14.3) that, for water, p shows a dependence on T which is closely exponential, at least over limited ranges of temperature. The variation of h_v with T is shown in Figure 14.7: h_v decreases as T increases and becomes zero when T equals T_c. Interestingly, Equation (14.15) predicts an exponential dependence of p on T provided that h_v is constant — a condition that Figure 14.7 shows is never satisfied.

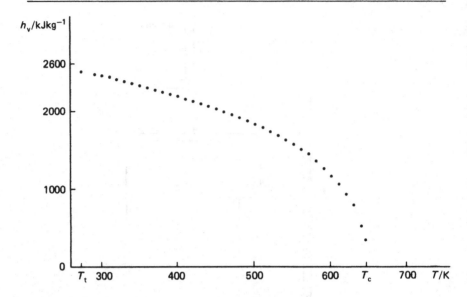

Figure 14.7 *The temperature dependence of the specific enthalpy of vaporisation* h_V *of water*

For many liquids that do not dissociate on vaporisation it is found that the molar enthalpy of vaporisation $H_{m,v}$ divided by the normal boiling point T_b is approximately constant and equal to about 85 J mol^{-1} K^{-1}. This result is usually known as Trouton's rule and is illustrated by the results in Table 14.1.

Table 14.1 Molar enthalpy of vaporisation $H_{m,v}$ and normal boiling temperature T_b for some common substances

Substance	$H_{m,v}$/kJ mol^{-1}	T_b/K	$H_{m,v} \div T_b$ /J mol^{-1} K^{-1}
Helium	0.08	4.22	19
Hydrogen	0.90	20.28	44.4
Nitrogen	5.59	77.35	72.3
Oxygen	6.82	90.19	75.6
Chlorine	20.41	239.1	85.4
Mercury	59.15	629.81	93.9
Caesium	65.90	958	68.8
Zinc	115.31	1180	97.7
Lead	179.41	2023	88.6

Values from Kaye and Laby (1973).

There is no corresponding general relation for the fusion of simple solids. However, when the molar enthalpy of fusion $H_{m,f}$ is divided by the normal melting point T_m for solids having the same crystal structure and type of bonding, an approximately constant value is obtained, as illustrated by the results in Table 14.2.

Table 14.2 Molar enthalpy of fusion $H_{m,f}$ for some substances having different crystal structures and bonding type

Substance		$H_{m,f}/kJ\ mol^{-1}$	T_m/K	$H_{m,f} \div T_m$ /J mol^{-1} K^{-1}
Sodium		2.60	371	7.00
Potassium	b.c.c.	2.32	336.4	6.90
Rubidium		2.34	312.0	7.50
Copper		13.05	1351.6	9.66
Silver	f.c.c.	11.30	1235.1	9.15
Aluminium		10.67	933.5	11.43
Neon		0.34	24.56	13.84
Argon	f.c.c.	1.18	83.75	14.00
Krypton		1.64	115.9	14.15

b.c.c. = body-centred cubic structure; f.c.c. = face-centred cubic structure.
Values from Kaye and Laby (1973).

■ EXAMPLE

Q. A mass of 0.025 kg of liquid helium at its boiling point (4.0 K under the pressure of the experiment) is contained in a Dewar vessel whose thermal capacity may be neglected. Calculate the percentage of the liquid that is evaporated when a small silver sphere of mass 0.350 kg and at a temperature 20.0 K is placed in the liquid helium. The specific enthalpy of vaporisation of liquid helium at 4 K is 2.1×10^4 J kg^{-1} and below 20 K the specific heat capacity c_p of silver varies with temperature T according to the equation

$$c_p/J\ kg^{-1}\ K^{-1} = 1.5 \times 10^{-4}\ (T/K)^3 + 6.0 \times 10^{-3}\ T/K$$

A. Let the mass of the sphere be M and that of the liquid helium be m.

When the sphere absorbs an infinitesimal quantity of heat q_s, it suffers a temperature change dT, given by

$$q_s = Mc_p dT$$

When the liquid helium absorbs an infinitesimal quantity of heat q_h the mass of liquid vaporised is dm, where

$$q_h = -h_v dm$$

and h_v is the specific enthalpy of vaporisation.

A negative sign is introduced here, since the absorption of a positive quantity of heat reduces the mass in the liquid phase.

Regarding the liquid helium and the sphere as an isolated system, in the process considered the heat absorbed by the sphere Q_s plus that absorbed by the helium Q_h is zero — that is,

$$Q_s + Q_h = 0$$

or

$$Q_h = - Q_s$$

The heat absorbed by the sphere is given by

$$Q_s = M \int_{T_i}^{T_f} c_p \mathrm{d}T$$

where T_i and T_f are, respectively, the initial and final temperatures of the sphere. Now Q_h is given by

$$Q_h = - h_v \int_{m_i}^{m_f} \mathrm{d}m$$

where m_i and m_f are, respectively, the initial and final masses of the helium in the liquid phase.

Therefore,

$$h_v \int_{m_i}^{m_f} \mathrm{d}m = M \int_{T_i}^{T_f} c_p \mathrm{d}T$$

and

$$2.1 \times 10^4 (m_f - m_i) = 0.35 \int_{20}^{4} (1.5 \times 10^{-4} T^3 + 6 \times 10^{-3} T) \mathrm{d}T$$

$$= 0.35 \left[1.5 \times 10^{-4} \frac{T^4}{4} + 6 \times 10^{-3} \frac{T^2}{2} \right]_{20}^{4}$$

$$= - 24998.4 \times 10^{-4}$$

Therefore,

$$m_f - m_i = - \frac{24\,998.4 \times 10^{-4}}{2.1 \times 10^4}$$

$$= - 1.19 \times 10^{-4} \mathrm{kg} = \Delta m$$

and

$$\frac{|\Delta m|}{m_i} = 0.004\ 76 \sim 0.48\%$$

■ EXERCISES

1   When a first-order phase change takes place in a closed hydrostatic system, part of the energy absorbed by the system is used to increase the internal energy and part to do work because of the change in volume. Using the values for water, show that the work done on the surroundings forms only a small part of the enthalpy of vaporisation. For water the specific enthalpy of vaporisation is 2257 kJ kg^{-1} at a pressure of 0.101 MPa; the specific volume of liquid water is 1.044×10^{-3} m^3 kg^{-1} and that of steam is 1673.9×10^{-3} m^3 kg^{-1}.

2   The saturation vapour pressure p of ammonia is given by

$$\ln p = 27.92 - \frac{3754}{T}$$

along the sublimation curve and by

$$\ln p = 24.38 - \frac{3063}{T}$$

along the vaporisation curve, p being in Pa in both equations. Calculate

(a) the triple-point temperature for ammonia,
(b) the molar enthalpy of vaporisation at the triple point.

3   The specific enthalpy of vaporisation h_v for water varies with temperature T as shown. Calculate the specific heat capacity of saturated steam at a temperature of 630 K if the specific heat capacity of saturated liquid water at that temperature is 13.3 kJ kg^{-1} K^{-1}.

T/K	620.5	623.0	625.4	627.8	630.1	632.4	634.6	636.8
h_v/kJ kg^{-1}	931.8	896.6	860.2	822.2	782.1	738.9	691.4	638.7

4   Calculate the minimum amount of work that must be supplied to a refrigerator to convert 1 kg of water, initially at 20 °C and atmospheric pressure, to ice at 0 °C under the same pressure, if the temperature of the surroundings is 20 °C. The specific heat capacity of liquid water at constant pressure may be taken as a constant equal to 4200 J kg^{-1} K^{-1} and the specific enthalpy of fusion of ice at atmospheric pressure as 34×10^4 J kg^{-1}.

5   At 0 °C the density of ice is 917.4 kg m^{-3}, while that of liquid water is
    1.00 × 10^3 kg m^{-3}, and the specific enthalpy of fusion is 3.35 × 10^5 J kg^{-1}.
    Use these values to determine the slope of the fusion curve for water at
    0 °C. Discuss the relevance of the result to ice-skating, remembering that
    the bottom of an ice skate is hollow-ground.

6   The density of water at a temperature of 100 °C and a pressure of 1
    (standard) atm is 958.8 kg m^{-3}, while the corresponding value for steam is
    0.597 kg m^{-3}. If the specific enthalpy of vaporisation under these conditions
    is 2257 kJ kg^{-1}, calculate the slope of the vaporisation curve at a
    temperature of 100 °C and a pressure of 1 atm. Estimate the boiling point
    of liquid water at the top of Mount Everest, where the pressure is 0.35
    standard atmosphere.

7   Show that the change in internal energy ΔU of a one-component closed
    hydrostatic system when it undergoes a first-order phase change at a
    temperature T under a pressure p is given by

$$\Delta U = H_{12}\left(1 - \frac{p}{T}\frac{dT}{dp}\right)$$

    where H_{12} is the total enthalpy change of the transition.

8   Show that, for a closed hydrostatic system consisting of two phases
    coexisting in equilibrium at a temperature T and under a pressure p,

$$\left(\frac{\partial p}{\partial V}\right)_S = -\frac{T}{C_V}\left(\frac{dp}{dT}\right)^2$$

    where C_V is the heat capacity of the system at constant volume and dp/dT
    is the slope of the appropriate phase equilibrium curve.

Notes

1   The term 'phase' has previously been used rather loosely (e.g. Section
    2.1) to mean a different condition of aggregation of the matter
    comprising a system. A more precise definition is that a phase is a
    homogeneous part of a system, bounded by surfaces across which
    some properties change discontinuously.

2   So called because the absorption of heat by a system during an
    ordinary phase change does not produce a change in temperature.

3   The normal boiling point of a liquid is that temperature at which the
    saturation vapour pressure is 1 standard atmosphere. It does not differ
    by a measurable amount from the temperature at which the liquid boils
    freely in dry air under standard atmospheric pressure.

15

The third law of thermodynamics

The third law of thermodynamics was proposed by Nernst as his Heat Theorem in 1906, following a study of chemical equilibrium. A simple statement of the third law may be obtained from a consideration of the Gibbs–Helmholtz equation (see Exercise 1 in Chapter 10), which, for a closed hydrostatic system, may be written

$$F = U + T\left(\frac{\partial F}{\partial T}\right)_V \tag{15.1}$$

where F is the Helmholtz function of the system and U its internal energy at a temperature T. Equation (15.1) predicts that, as T tends to absolute zero, the value of F approaches the value of U. It is also clear from Equation (15.1) that, when F is known, U may be calculated. However, knowledge of U does not allow F to be uniquely determined from Equation (15.1), since any solution of that equation, plus any term of the form (constant $\cdot T$) is also a solution. This may be seen from the integrated form of Equation (15.1). Since

$$\left(\frac{\partial(F/T)}{\partial T}\right)_V = \frac{1}{T}\left(\frac{\partial F}{\partial T}\right)_V - \frac{F}{T^2}$$

$$F/T = -\int U/T^2 \, dT + \text{function of integration} \tag{15.2}$$

Nernst guessed that the actual solution could be found from the behaviour of the system as its temperature approached absolute zero. He noted that, in naturally occurring chemical reactions, the change in U is often not very different from the change in F, especially at not too high temperatures, and postulated that, in fact, the curves of ΔU and ΔF run together asymptotically at absolute zero — that is,

$$\lim_{T\to 0} \frac{d\Delta F}{dT} = \lim_{T\to 0} \frac{d\Delta U}{dT} = 0 \tag{15.3}$$

Total derivations are used here, since the result should be true whether pressure or volume is held constant.

Equation (15.3), which is the earliest statement of the third law, enables a unique solution of Equation (15.2) to be obtained when either ΔF or ΔU is known.

Some applications of this result produced examples where the rule was found not to hold, but careful investigation showed that the apparent violations occurred in systems that were not in thermodynamic equilibrium. The theorem was put on a sound basis largely by the work of F. E. Simon, and it is now established as the third law of thermodynamics. In this chapter a simple modern statement of the third law is given and some straightforward applications are considered.

15.1 The third law

When a closed hydrostatic system undergoes an infinitesimal reversible process at constant pressure, the change in entropy dS is given by

$$dS = C_p \frac{dT}{T} \tag{15.4}$$

where C_p is the heat capacity at constant pressure when the temperature is T. The most adequate way of making Equation (15.4) satisfy relation (15.3) is by making C_p tend to zero as T tends to zero.

Experimentally it is found that C_p approaches zero as the temperature approaches absolute zero, so that Equation (15.4) may be integrated from absolute zero to any arbitrarily chosen temperature T_1 without there being a singularity in the entropy as the temperature approaches zero. Integrating Equation (15.4) then gives

$$S_{T_1} = S_0 + \int_0^{T_1} \frac{C_p}{T} dT \tag{15.5}$$

where S_0 is the value of the entropy at absolute zero and the integral is finite.

Equation (15.5) gives no information about S_0 and, in particular, does not indicate whether S_0 has the same value if the system is held at a different constant pressure.

Now, by definition, the Helmholtz function F is

$$F = U - TS$$

so that, in a process that takes place at constant temperature,

$$\Delta F = \Delta U - T \Delta S \tag{15.6}$$

To find out how $T\Delta S$ varies as the temperature of the process varies, it is necessary to examine $\mathrm{d}(\Delta F)/\mathrm{d}T$. From Equation (15.6) this is given by

$$\frac{\mathrm{d}(\Delta F)}{\mathrm{d}T} = \frac{\mathrm{d}(\Delta U)}{\mathrm{d}T} - T\frac{\mathrm{d}(\Delta S)}{\mathrm{d}T} - \Delta S \qquad (15.7)$$

Applying Nernst's postulate (Equation 15.3) to Equation (15.7) gives the result: in any reversible isothermal process the entropy change ΔS tends to zero as the temperature tends to zero. This statement will be called the third law of thermodynamics. The term 'reversible' implies that the system must be in thermodynamic equilibrium. The third law states that all curves of S against T must run together at absolute zero, as in Figure 15.1. From this it follows that the value of S_0 is independent of other variables.

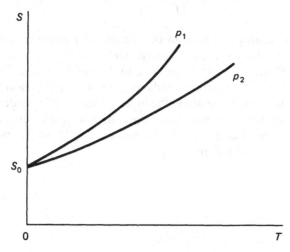

Figure 15.1 *Entropy curves for a closed hydrostatic system in the neighbourhood of absolute zero*

The vanishing of the heat capacities as the temperature approaches absolute zero allows absolute zero to be used as a convenient reference level for entropy calculations. Putting S_0 equal to zero in Equation (15.5), for all simple substances, allows a unique value to be given to the entropy at any other temperature.

Since classical thermodynamics is concerned with the change in entropy of a system in two given equilibrium states, the actual value assigned to S_0 is immaterial. In fact, S_0 cannot be evaluated on the basis of classical thermodynamics. However, from the statistical mechanics of weakly coupled systems [1], it appears that the entropy of a system that is in thermodynamic equilibrium does approach zero as the temperature approaches zero. This statement is useful for pure crystals, but not for alloys and glasses, where diffusion processes are so slow at low temperatures that the system is always in a state of metastable equilibrium and the simple statement of the third law is not applicable.

15.2 Some applications of the third law

A striking confirmation of the third law of thermodynamics is given by the melting of helium. The dependence of melting pressure p on temperature T is given by the Clapeyron–Clausius equation, which may be written in the form

$$\frac{\mathrm{d}p}{\mathrm{d}T} = \frac{\Delta S}{\Delta V} \tag{15.8}$$

where ΔS is the entropy change in the (reversible) phase transition (see Section 14.2). By the third law ΔS must tend to zero as the temperature tends to zero, which means that the slope of the melting curve $\mathrm{d}p/\mathrm{d}T$ must tend to zero as the temperature tends to zero. This prediction has been tested for ^{4}He, which can remain in the liquid phase practically down to the absolute zero of temperature. The experimental results are shown in Figure 15.2. It is found that, for ^{4}He, the value of $\mathrm{d}p/\mathrm{d}T$ below about 1 K is practically temperature-independent.

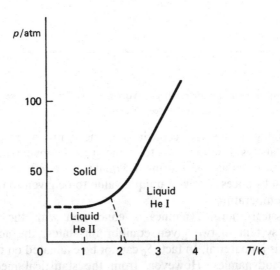

Figure 15.2 *The phase diagram for ^{4}He in the neighbourhood of absolute zero. From page 46 of* Third Law of Thermodynamics, *by J. Wilks, Oxford University Press, Oxford, 1961*

The third law is also able to predict the behaviour of the cubic expansivity β in the neighbourhood of absolute zero. For a closed hydrostatic system having a volume V at a temperature T under a pressure p, β is defined by

$$\beta = \frac{1}{V}\left(\frac{\partial V}{\partial T}\right)_p \tag{15.9}$$

Using Maxwell relation (M1) (page 114), Equation (15.8) may be written

$$\beta = -\frac{1}{V}\left(\frac{\partial S}{\partial p}\right)_T \tag{15.10}$$

By the third law $(\partial S/\partial p)_T$ tends to zero as T tends to zero, so that

$$\lim_{T\to 0}\beta = 0 \tag{15.11}$$

This result is in agreement with the result established by Gruneissen that the cubic expansivity is proportional to the specific heat capacity at constant pressure.

For a linear, isotropic, homogeneous paramagnetic solid the change in entropy with change in applied magnetic B-field B_a, is given by (see Equation 12.48)

$$\left(\frac{\partial S}{\partial B_a}\right)_T = \left(\frac{\partial m}{\partial T}\right)_{B_a} \tag{15.12}$$

where m is the magnetic moment of the sample. In terms of the susceptibility χ_m, Equation (15.12) may be written

$$\left(\frac{\partial S}{\partial B_a}\right)_T = \frac{VB_a}{\mu_0}\left(\frac{\partial \chi_m}{\partial T}\right)_{B_a} \tag{15.13}$$

so that, by the third law,

$$\lim_{T\to 0}\left(\frac{\partial \chi_m}{\partial T}\right)_{B_a} = 0 \tag{15.14}$$

Equation (15.14) shows that Curie's law, $\chi_m = a/T$, cannot hold at all temperatures down to absolute zero, since $(\partial \chi_m/\partial T)_{B_a}$ equals $-a/T^2$, which does not vanish as T tends to zero. Equation (15.13) indicates that, as T tends to zero, χ_m must become independent of temperature, so that $(\partial \chi_m/\partial T)_{B_a}$ becomes zero.

15.3 The unattainability of absolute zero

The temperature of any system may be reduced by operating a refrigerator between it and a reservoir at a higher temperature. This procedure is most efficient when the refrigerator operates in a reversible manner and the lowest temperature that may be attained is that at which the heat influx to the system from the surroundings in one cycle of operation equals the heat removed per cycle.

Low temperatures may also be achieved using non-cyclic processes, which, for maximum efficiency, should also be carried out reversibly. Provided that the entropy of the system is a decreasing function of the temperature and is also dependent on some other independent coordinate (for example, pressure), a temperature drop is produced by a suitable isentropic process, as indicated in Figure 15.3, where the pressure p is changed from p_1 to p_2. Starting from a state i, in which the temperature is T_i, the isentropic process ends in a state f in which the temperature is T_f. Ever lower temperatures can then be produced by allowing the system to undergo consecutive processes that are alternately reversible isothermals and reversible adiabatics, limited by the entropy curves corresponding to the independent coordinate values p_1 and p_2. However, according to the

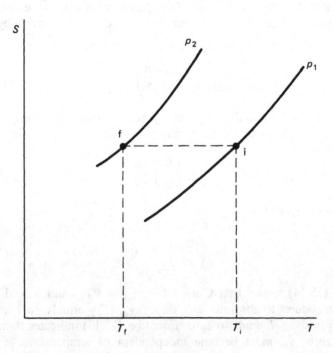

Figure 15.3 *The production of low temperatures by a reversible adiabatic process*

third law, entropy changes in reversible isothermal processes become progressively smaller as absolute zero is approached, so that the entropy curves for different values of the other independent coordinates converge as absolute zero is approached, as in Figure 15.4. This makes the reaching of absolute zero impossible, except as a limiting process with an infinite number of steps. This result is often called the unattainability statement of the third law: it is impossible to reduce the temperature of a system to the absolute zero in a finite number of operations.

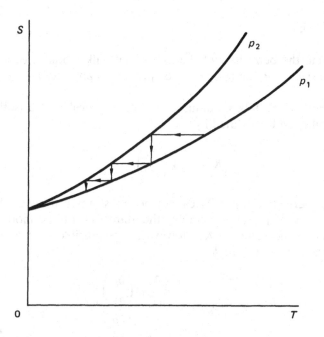

Figure 15.4 *To illustrate that absolute zero is unattainable in a finite number of steps*

The same result may also be obtained by considering the second $T\mathrm{d}S$ equation (see Exercise 9 in Chapter 11), which is, for a closed hydrostatic system,

$$T\mathrm{d}S = C_p\mathrm{d}T - T\left(\frac{\partial V}{\partial T}\right)_p \mathrm{d}p \qquad (15.15)$$

In an isentropic process $\mathrm{d}S$ is zero, so that the temperature change $\mathrm{d}T_S$ is given by

$$\mathrm{d}T_S = \frac{T}{C_p}\left(\frac{\partial V}{\partial T}\right)_p \mathrm{d}p$$

Using Maxwell relation (M1) (page 114), this becomes

$$dT_S = - \frac{T}{C_p} \left(\frac{\partial S}{\partial p} \right)_T dp \qquad (15.16)$$

According to the third law, $(\partial S/\partial p)_T$ tends to zero as T tends to zero, so that dT_S becomes progressively smaller as T is reduced — that is, the absolute zero cannot be attained from any finite temperature T.

■ EXAMPLE

Q. Examine the behaviour of the isothermal bulk modulus of a closed hydrostatic system as the temperature approaches absolute zero.

A. The isothermal bulk modulus K is the reciprocal of the isothermal compressibility and is defined by

$$K = - V \left(\frac{\partial p}{\partial V} \right)_T \qquad (15.17)$$

The partial derivative $(\partial p/\partial V)_T$ cannot be transformed using a Maxwell relation, so that, not unexpectedly, the third law of thermodynamics cannot predict the value of K. However, it is possible to examine the temperature dependence of K.

$$\left(\frac{\partial K}{\partial T} \right)_V = - V \frac{\partial}{\partial T} \bigg)_V \left(\frac{\partial p}{\partial V} \right)_T$$
$$= - V \frac{\partial}{\partial V} \bigg)_T \left(\frac{\partial p}{\partial T} \right)_V \qquad (15.18)$$

Now

$$\left(\frac{\partial p}{\partial T} \right)_V = \left(\frac{\partial S}{\partial V} \right)_T \qquad (M3)$$

so that

$$\left(\frac{\partial K}{\partial T} \right)_V = - V \frac{\partial}{\partial V} \bigg)_T \left(\frac{\partial S}{\partial V} \right)_T \qquad (15.19)$$

According to the third law, the change in entropy in a reversible isothermal process tends to zero as the temperature tends to zero — that is,

$$\left(\frac{\partial S}{\partial V} \right)_T \to 0 \text{ as } T \to 0$$

Therefore,

$$\lim_{T\to 0}\left(\frac{\partial K}{\partial T}\right)_V = 0$$

or, as the absolute zero of temperature is approached the value of the bulk modulus becomes independent of temperature.

■ EXERCISES

1   Use the third law of thermodynamics to obtain the following results.

(a) For a saturated reversible voltaic cell of e.m.f. E at a temperature T

$$\lim_{T\to 0}\frac{dE}{dT} = 0$$

(b) For the surface of a liquid in equilibrium with its own vapour, having a specific surface free energy σ at a temperature T,

$$\lim_{T\to 0}\frac{d\sigma}{dT} = 0$$

2   Discuss the validity of the following argument, due to Nernst, which endeavours to show that the third law of thermodynamics is a consequence of the second law.

Carry out a Carnot cycle between, say, a reservoir at room temperature and a finite body at a lower temperature. If the lower temperature is absolute zero, there is no heat transfer in the corresponding reversible isothermal process and all the heat supplied at the high temperature is converted into work. This, however, contravenes Kelvin's statement of the second law. Therefore, the absolute zero is unattainable.

3   Show that the pressure coefficient of a closed hydrostatic system, $(\partial p/\partial T)_V$, tends to zero as the temperature tends to absolute zero.

Use this result to show that van der Waals' equation cannot be valid at low temperatures.

What does the third law say about the existence of the classical ideal gas?

Note

1   See, for example, Chapter 4 in *Statistical Physics*, by F. Mandl (London: Wiley, 1971).

16

The application of thermodynamics to some irreversible processes

Each equilibrium state of a system is characterised by a set of values of the thermodynamic coordinates which do not depend on the way in which the equilibrium state was reached. Processes that are, effectively, a succession of equilibrium states can also be described in terms of the coordinates, but this is not true of processes in which there are finite gradients in one or more of the intensive coordinates. Such processes are, of course, irreversible. However, provided that a given process takes place between an initial equilibrium state and a final equilibrium state, a thermodynamical treatment can be used to relate the values of the coordinates in those equilibrium states, even though it can give no information about the process. The technique is to find a notional reversible process linking the required initial and final equilibrium states and then use that process to calculate relations between the thermodynamic coordinates (or state functions). This procedure will now be examined in detail as it is applied to two irreversible processes.

16.1 The Joule effect

One irreversible process linking two equilibrium states which has already received some discussion is the free expansion of a gas (see Section 6.2). In a free expansion, a gas is allowed to expand in such a way that there is no thermal or work interaction with the surroundings. It follows from the first law that the gas then suffers no change in internal energy. A free expansion is highly irreversible: finite pressure gradients are set up in the gas and bulk kinetic energy is given to the gas, this energy then being dissipated by viscous forces.

Let the initial equilibrium state i of the gas be characterised by a pressure p_i, volume V_i and temperature T_i, and let the respective values for the final equilibrium state f be p_f, V_f, T_f. The internal energy U has the same value in these two states, so that a convenient notional reversible process linking states i and f is one that takes place at constant U.

The change in temperature which occurs in an infinitesimal element of the notional reversible process will be determined by $(\partial T/\partial V)_U$, a quantity known as the *Joule coefficient*. Using a reciprocity relation, this coefficient may be written

$$\left(\frac{\partial T}{\partial V}\right)_U = -\left(\frac{\partial U}{\partial V}\right)_T\left(\frac{\partial T}{\partial U}\right)_V.$$ (16.1)

Now, for a reversible process,

$$\left(\frac{\partial U}{\partial T}\right)_V = C_V,$$

the heat capacity at constant volume, and

$$\left(\frac{\partial U}{\partial V}\right)_T = T\left(\frac{\partial p}{\partial T}\right)_V - p \text{ (the energy equation)}$$

Substitution in Equation (16.1) gives

$$\left(\frac{\partial T}{\partial V}\right)_U = -\frac{1}{C_V}\left[T\left(\frac{\partial p}{\partial T}\right)_V - p\right]$$ (16.2)

It will be noted that an expression for $(\partial T/\partial V)_U$ has been derived without the need to devise an experimental arrangement to bring about a reversible process at constant internal energy. In a finite process, in which the volume of the gas changes from V_i to V_f, the change in temperature is ΔT, given by

$$\Delta T = T_f - T_i = -\int_{V_i}^{V_f} \frac{1}{C_V}\left[T\left(\frac{\partial p}{\partial T}\right)_V - p\right]dV$$ (16.3)

Note that Equations (16.2) and (16.3) contain only one principal heat capacity and quantities that may be obtained from the equation of state of the gas.

For an ideal gas the integrand in Equation (16.3) is always zero and so a free expansion produces no change in temperature. Real gases, however, always undergo a fall in temperature as a result of free expansion. This can be seen from Equation (16.1), since both $(\partial T/\partial U)_V$ and $(\partial U/\partial V)_T$ are always positive: the first term is an inverse heat capacity and the second represents the additional energy needed to increase the molecular separation against the molecular attractions. This will be discussed further in Section 17.2.

To effect the integration of Equation (16.3) the equation of state of the gas and the volume-dependence of C_V must both be known. The volume-dependence of C_V is given by Equation (11.16), which also requires a knowledge of the equation of state. However, for an ideal gas C_V is independent of V and this is likely to be a reasonable approximation for dilute real gases. It is convenient in this context to take the equation of state of a real gas in series form. When the gas is dilute, the equation may then be written

$$pV_m = RT + Bp \qquad (16.4)$$

where V_m is the molar volume, R is the gas constant and B is a function of temperature only. It is more useful to have pV_m as a function of V_m, and this can be achieved by using the approximate result for dilute real gases that

$$pV_m = RT + \frac{BRT}{V_m} \qquad (16.5)$$

From Equation (16.5)

$$\left(\frac{\partial p}{\partial T}\right)_V = \frac{R}{V_m} + \frac{R}{V_m^2}\left[B + T\frac{dB}{dT}\right]$$

and substituting in Equation (16.3) gives

$$T_f - T_i = -\frac{1}{C_V}\int_{V_i}^{V_f}\left(\frac{RT}{V_m} + \frac{RT}{V_m^2}\left[B + T\frac{dB}{dT}\right] - p\right)dV \qquad (16.6)$$

Using Equation (16.5) to obtain an expression for p, Equation (16.6) gives

$$T_f - T_i = -\frac{1}{C_V}\int_{V_i}^{V_f}\left(\frac{RT}{V_m} + \frac{RT}{V_m^2}\left[B + T\frac{dB}{dT}\right] - \frac{RT}{V_m} - \frac{BRT}{V_m^2}\right)dV$$

$$= -\frac{1}{C_V}\int_{V_i}^{V_f}\frac{RT^2}{V_m^2}\frac{dB}{dT}dV \qquad (16.7)$$

Consider the effect of allowing 1 mol of molecules of nitrogen at s.t.p. to undergo a free expansion in which the volume is doubled — that is, $V_f = 2V_i$. For nitrogen at s.t.p. $C_{V,m} = 5R/2$ and $V_m = 22.4 \times 10^{-3}$ m³. The temperature-dependence of B has the same form for all gases; that for

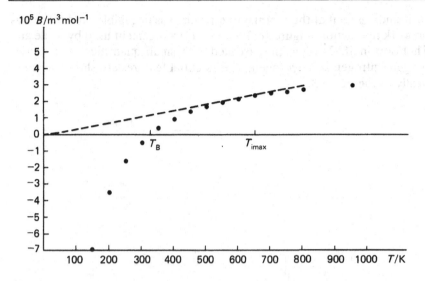

Figure 16.1 *The temperature dependence of the coefficient B for nitrogen*

nitrogen is shown in Figure 16.1. At a temperature of 273 K the value of dB/dT for nitrogen is $0.024 \times 10^{-5} \mathrm{~m}^3 \mathrm{~K}^{-1} \mathrm{~mol}^{-1}$. Assuming that $T_f - T_i$ is small, so that T and dB/dT in Equation (16.7) may be treated as constants, integration gives

$$T_f - T_i = \frac{RT^2}{C_V} \frac{dB}{dT} \left[\frac{1}{V_f} - \frac{1}{V_i} \right] \tag{16.8}$$

Substituting the values given makes $\Delta T = 0.16$ K, which is sufficiently small to justify the assumptions made, and explains the null result obtained by Joule in his experiments on the free expansion of air.

16.2 The Joule–Thomson effect

Joule's experiments on the free expansion of gases (1845) were intended to give information about intermolecular attractions, but suffered from being insensitive. This problem was overcome in a series of experiments using a continuous flow of gas, carried out, in collaboration with W. Thomson (later Lord Kelvin), in the years 1853–1862. In these experiments a sample of the gas was made to expand from a high pressure to a lower pressure by passage through a 'throttle', which is a plug of porous material (cotton wool or silk in the original experiments), although a small hole in a diaphragm produces the same effect. The apparatus was insulated from the surroundings so that the thermal interaction was negligible, and there was no work interaction. Figure 16.2 shows the arrangement used by Joule and Thomson in 1862, when they worked with small quantities of the gases oxygen, nitrogen and hydrogen. The essential features are shown schematically in Figure 16.3.

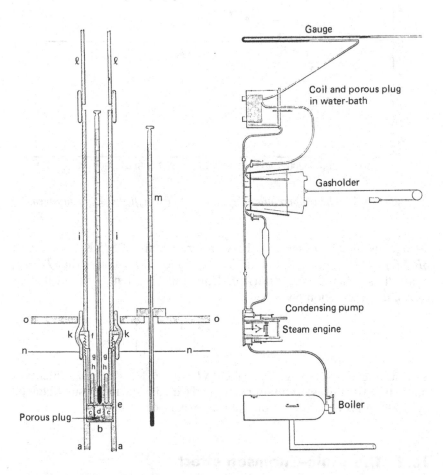

Figure 16.2 *The Joule–Thomson porous plug experiment. Based on pages 89 (Figure 1) and 91 (Figure 2) of* The Free Expansion of Gases, *translated and edited by J. S. Ames, Harper and Brothers, New York, 1898*

The Joule–Thomson process is a continuous-flow process. Pressure gradients are present in the gas and any convenient thermodynamic system, 1 mol of molecules of the gas, say, is not in thermodynamic equilibrium. However, when steady conditions are achieved, it is possible to make macroscopic measurements of the pressure and temperature of the gas at a particular location; these values characterise the state of the gas in

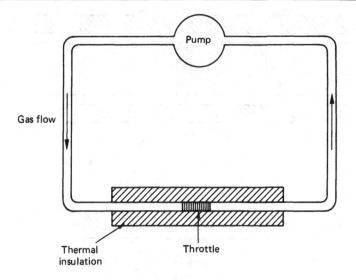

Figure 16.3 *The essential features of the Joule–Thomson experiment*

that region and do not vary with time. Therefore, the gas that is about to pass through the porous plug and that which has just passed through the plug may be treated as being in a state of *quasithermodynamic equilibrium* characterised by pressure and temperature p_i, T_i and p_f, T_f, respectively.

To determine the relationship between the coordinates in these two quasiequilibrium states, a model of the throttling process may be considered, as shown in Figure 16.4. The system is a fixed mass of gas at a temperature T_i and pressure p_i, maintained on the high-pressure side of the throttle by two pistons A and B (Figure 16.4a). These pistons reproduce the effect of the remainder of the flowing gas in the real arrangement of Figure 16.3. The pistons then move at appropriate, different but very slow, speeds so that the pressure of the gas between piston A and the throttle is maintained at p_i and that of the gas between the throttle and piston B is maintained at p_f. Both pistons are stopped when all the gas has just passed through the throttle and occupies a volume V_f at a pressure p_f (Figure 16.4b).

The throttling process itself is highly irreversible and cannot be described by thermodynamic coordinates, but it is possible to apply the first law to the equilibrium states shown in Figure 16.4. Let the internal energy of the mass of gas be U_i when it is in the state shown in Figure 16.4(a) and U_f when it is in the state shown in Figure 16.4(b). Since the system boundary is adiabatic, the thermal interaction with the surroundings is zero. Applying the first law to the process represented by Figure 16.4 gives

$$\Delta U = U_f - U_i = W_F$$

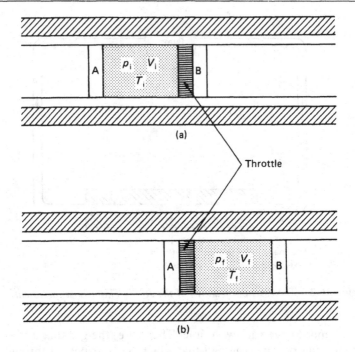

Figure 16.4 *A model for the throttling process*

where W_F is the work done on the system. In the arrangement of Figure 16.3 there is no work interaction with the surroundings. The equivalent of W_F in the model arrangement of Figure 16.4 is the work needed to maintain the steady gas flow, and for this reason W_F is known as the flow work.

Piston A does work on the system given by

$$- \int_{V_i}^{0} p_i \mathrm{d}V$$

which equals $p_i V_i$, since p_i remains constant during the process. Similarly, the work done by piston B is $- p_f V_f$. Therefore,

$$W_F = p_i V_i - p_f V_f$$

and the equation for ΔU becomes

$$U_f - U_i = p_i V_i - p_f V_f$$

or

$$U_f + p_f V_f = U_i + p_i V_i \qquad (16.9)$$

Since the enthalpy H is given by $H = U + pV$, this result may be written

$$H_f = H_i \tag{16.10}$$

Equation (16.10) shows that the enthalpy of a given mass of gas is the same before and after a throttling process. It does not mean that the enthalpy remains constant during the process; the process is irreversible and cannot be described by thermodynamic coordinates.

A typical arrangement for the study of the throttling process, used by Hoxton in 1919, is shown in Figure 16.5. The gas passes through a constant temperature bath and its temperature T_i is measured using the platinum resistance thermometer A. Then the gas enters the porous plug casing at B. The plug P is in the form of a hollow earthenware cylinder closed at one end and divided into two parts by the guard ring GG. That part of the gas stream passing through the plug below GG is guided by the tube tt past the resistance thermometer C, where the final steady temperature T_f is measured. A radiation shield SS minimises the thermal interaction between the gas and the surroundings and the guard ring GG prevents effects caused by heat transfer from the surroundings along the plug itself.

The usual procedure is to select values for p_i and T_i, since these may be chosen at will, set p_f to a suitable value and measure the resulting value of

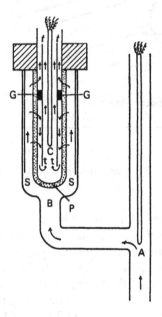

Figure 16.5 *The apparatus used by Hoxton to study the Joule–Thomson effect. Taken from page 374 (Figure 143) of* A Text-book of Heat: Part I, *by H. S. Allen and R. S. Maxwell, Macmillan, London, 1959*

T_f. The enthalpy of any fixed mass of gas — for example, 1 mol — has the same value in state i and state f. Therefore, by keeping p_i and T_i fixed and taking different values of p_f, a series of states having the same enthalpy are obtained. The line joining the points on a p–T graph that represent these states is a line of constant enthalpy or *isenthalp*. By choosing initial states of different enthalpy a series of isenthalpic curves is obtained.

The lines of constant enthalpy that are obtained for real permanent gases have the form shown in Figure 16.6. Provided that the initial temperature is low enough, the isenthalps pass through a maximum when plotted in the p–T plane. At this maximum $(\partial T/\partial p)_H$, which is known as the

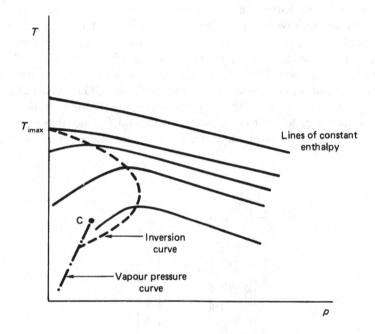

Figure 16.6 *The isenthalps of a throttling process.* C = *critical point*

Joule–Thomson coefficient μ, is zero. The locus of points for which μ is zero is known as the *inversion curve* and the temperature at which μ is zero is the *inversion temperature* T_i for the particular isenthalp. There is a maximum value for the inversion temperature T_{imax}, given by the value at which the inversion curve cuts the temperature axis, which is characteristic of the particular gas. Values of T_{imax} for some gases that have been used in producing low temperatures are given in Table 16.1.

Whether $T_f > T_i$ or $T_f < T_i$ depends on the initial and final states of the throttling process, but T_f is always greater than T_i when T_i is greater than T_{imax}.

Table 16.1 Values of the critical temperature T_c, triple-point temperature T_t and approximate maximum inversion temperature T_{imax} for some gases that have been used in the production of low temperatures

Substance	T_c/K	T_t/K	T_{imax}/K
Methly chloride	416.3		
Ammonia	405.4	195.42	
Carbon dioxide	304.2	216.55	1500
Ethylene	282.4	104.00	
Krypton	209.4	115.6	1089
Methane	190.6	90.67	970
Oxygen	154.8	54.35	890
Argon	150.7	83.78	720
Nitrogen	126.2	63.15	620
Neon	44.4	24.57	210
Hydrogen	32.99	13.84	200
Helium (^{4}He)	5.2		50

Values of T_c from Kaye and Laby (1973).
Values of T_t from Hsieh (1975).

To obtain a relationship between the primitive coordinates in the initial and final equilibrium states, the irreversible process must be replaced by a notional reversible process linking the same two states. In view of the previous analysis, it is convenient to choose a notional reversible process that is carried out at constant enthalpy.

Consider a fixed mass of gas undergoing a reversible process at constant enthalpy. Any change in temperature may be obtained from the Joule–Thomson coefficient μ, which may be written

$$\mu = \left(\frac{\partial T}{\partial p}\right)_H = -\left(\frac{\partial T}{\partial H}\right)_p \left(\frac{\partial H}{\partial p}\right)_T \qquad (16.11)$$

by using a reciprocity relation.

Now, by definition,

$$H = U + pV$$

where V is the volume occupied by the fixed mass of gas. Therefore,

$$dH = dU + pdV + Vdp$$

and, since

$$dU = TdS - pdV \quad \text{(first law)}$$

$$dH = TdS + Vdp \qquad (16.12)$$

Equation (16.12) can be used to obtain expressions for both $(\partial T/\partial H)_p$ and $(\partial H/\partial p)_T$ — namely,

$$\left(\frac{\partial H}{\partial T}\right)_p = T\left(\frac{\partial S}{\partial T}\right)_p = C_p \qquad (16.13)$$

and

$$\left(\frac{\partial H}{\partial p}\right)_T = T\left(\frac{\partial S}{\partial p}\right)_T + V \qquad (16.14)$$

Using relation (M1) (page 114), Equation (16.14) becomes

$$\left(\frac{\partial H}{\partial p}\right)_T = V - T\left(\frac{\partial V}{\partial T}\right)_p \qquad (16.15)$$

Substituting from Equations (16.13) and (16.15) into Equation (16.11) gives

$$\mu = \left(\frac{\partial T}{\partial p}\right)_H = \frac{1}{C_p}\left[T\left(\frac{\partial V}{\partial T}\right)_p - V\right] \qquad (16.16)$$

$$= \frac{V}{C_p}(\beta T - 1) \qquad (16.17)$$

where β is the cubic expansivity (at constant pressure). Along the inversion curve μ is zero, so that the equation of the inversion curve may be written

$$T_i = \frac{1}{\beta} \qquad (16.18)$$

Note that Equation (16.16) contains only a principal heat capacity and quantities that may be obtained from the equation of state.

In a throttling process involving a finite pressure drop

$$T_f - T_i = \int_{p_i}^{p_f} \frac{1}{C_p}\left[T\left(\frac{\partial V}{\partial T}\right)_p - V\right]dp \qquad (16.19)$$

For a system consisting of n moles of molecules of ideal gas

$$pV = nRT$$

so that

$$T\left(\frac{\partial V}{\partial T}\right)_p - V = \frac{TnR}{p} - V = 0$$

Therefore, a throttling process produces no change in the temperature of an ideal gas, whatever the conditions.

The behaviour of real gases may be examined using the equation of state in the series form

$$pV_m = RT + Bp + Cp^2 + \cdots \tag{16.20}$$

where B, C . . . are functions of temperature only. Then,

$$\left(\frac{\partial V}{\partial T}\right)_p = \frac{R}{p} + \frac{dB}{dT} + p\frac{dC}{dT} + \cdots$$

so that

$$\mu = \frac{1}{C_{p,m}}\left[\left(T\frac{dB}{dT} - B\right) + p\left(T\frac{dC}{dT} - C\right) + p^2\left(T\frac{dD}{dT} - D\right) + \cdots\right] \tag{16.21}$$

In the limit, as p becomes vanishingly small, the value of μ tends to the value μ_0, where

$$\mu_0 = \frac{1}{C_{p_0,m}}\left(T\frac{dB}{dT} - B\right) \tag{16.22}$$

and $C_{p_0,m}$ is the value of $C_{p,m}$ at vanishingly small pressure. This extrapolation is esentially from a region where $pV_m = RT + Bp$, which is why, even for p equal to zero, where the gas behaviour is ideal, a non-zero value of μ is obtained, dependent on an imperfection parameter.

The value of T_{imax} is the value of T_i when p is zero, and is found for a gas whose equation of state is Equation (16.20) by putting μ_0 equal to zero in Equation (16.22). Then

$$T = T_{imax}$$

when

$$\frac{dB}{dT} = \frac{B}{T} \tag{16.23}$$

For real permanent gases B varies with temperature as shown in Figure 16.1. When B is zero, the gas behaves in an almost ideal manner over a wide range of pressure and the corresponding temperature is the Boyle temperature T_B. Equation (16.23) shows that the maximum inversion temperature T_{imax} may be found by drawing the tangent to the curve of B against T which also passes through the origin. This is usually a better way of determining T_{imax} than doing it directly. It is clear from Figure 16.1 that T_{imax} is always greater than T_B.

The Joule–Thomson coefficient discussed above is strictly the isenthalpic coefficient. A less commonly used coefficient is the isothermal Joule–Thomson coefficient, φ, defined by

$$\varphi = \left(\frac{\partial H}{\partial p}\right)_T \tag{16.24}$$

Since H is a state function, it may be treated as a function of T and p. Then

$$dH = \left(\frac{\partial H}{\partial T}\right)_p dT + \left(\frac{\partial H}{\partial p}\right)_T dp = C_p dT + \left(\frac{\partial H}{\partial p}\right)_T dp$$

When the enthalpy is maintained constant in a process, dH is zero and

$$C_p \left(\frac{\partial T}{\partial p}\right)_H = - \left(\frac{\partial H}{\partial p}\right)_T$$

But

$$\left(\frac{\partial T}{\partial p}\right)_H = \mu$$

so that

$$\mu = - \frac{\varphi}{C_p} \tag{16.25}$$

φ may be determined experimentally by performing the usual isenthalpic expansion and then transferring energy to or from the gas until $T_i = T_f$.

16.3 Gas thermometer corrections

The Joule–Thomson effect may be used to correct gas thermometer readings for non-ideality of the gas. Measurements of the Joule–Thomson effect may be made using any arbitrary thermometer, which may include a gas thermometer. Let θ be the temperature measured by this thermometer, corresponding to a thermodynamic temperature T. If the pressure drop Δp across the throttle is not too large, Equation (16.19) may be written

$$\Delta T = \frac{1}{C_p}\left[T\left(\frac{\partial V}{\partial T}\right)_p - V\right]\Delta p \tag{16.26}$$

Since T must be some function of θ,

$$\Delta T = \frac{dT}{d\theta} \cdot \Delta\theta \tag{16.27}$$

where $\Delta\theta$ is the temperature change measured by the arbitrary thermometer. Then

$$\left(\frac{\partial V}{\partial T}\right)_p = \left(\frac{\partial V}{\partial\theta}\right)_p \cdot \frac{d\theta}{dT} \tag{16.28}$$

and

$$C_p = \left(\frac{q}{dT}\right)_p = \left(\frac{q}{d\theta}\right)_p \cdot \frac{d\theta}{dT} = C_p'\frac{d\theta}{dT} \tag{16.29}$$

where C_p' is the heat capacity at constant pressure determined using the arbitrary thermometer. Substituting from Equations (16.27) – (16.29) into Equation (16.26) gives

$$\frac{dT}{d\theta}\Delta\theta = \frac{T\left(\dfrac{\partial V}{\partial\theta}\right)_p \dfrac{d\theta}{dT} - V}{C_p' \dfrac{d\theta}{dT}} \cdot \Delta p$$

or

$$\frac{dT}{T} = \frac{\left(\dfrac{\partial V}{\partial\theta}\right)_p d\theta}{V + C_p' \dfrac{\Delta\theta}{\Delta p}}$$

At the triple point of water, $\theta = \theta_3$ and $T = T_3 = 273.15$ K by definition. Therefore,

$$\ln\frac{T}{T_3} = \int_{\theta_3}^{\theta} \frac{(\partial V/\partial\theta)_p}{V + C_p' \dfrac{\Delta\theta}{\Delta p}} d\theta \tag{16.30}$$

Equation (16.30) may be integrated, since it contains only quantities that are readily measured.

16.4 The liquefaction of gases

16.4.1 Liquefaction by pressure

The phase diagram for a pure substance (Figure 13.3) shows that the liquid phase exists only over a limited temperature range, bounded by the triple-point temperature T_t and the critical temperature T_c (only for helium does the liquid phase persist to the lowest temperatures). A vapour whose temperature lies between T_t and T_c may always be liquefied by the simple application of pressure (see Section 13.1). This technique was first used by Northmore in 1806 to liquefy chlorine and was subsequently used by Faraday, from 1823 onwards, to liquefy a number of gases, including chlorine, nitrous oxide and carbon dioxide. This method was also used by Wroblewski and Olszewski who, in 1883, succeeded in liquefying oxygen in bulk and obtained liquid oxygen boiling in a test tube. The necessary cooling was achieved by boiling liquid ethylene under reduced pressure.

Once liquid is produced, lower temperatures may be achieved by reducing the pressure over the liquid under adiabatic conditions. The removal of the vapour promotes further evaporation, the necessary enthalpy coming from the system, which then suffers a fall in temperature. This behaviour may be conveniently represented on the graph of entropy against temperature, which, for a pure substance, has the form shown in Figure 16.7. If the saturated liquid is in the state represented by A, the isentropic reduction of the pressure moves the point representing the state of the system along the line AE. This method can, in principle, be extended to the solid–vapour region, but, as the vapour pressure of the solid is small and falls off very rapidly as the temperature decreases, in this region it is not a very practical method.

16.4.2 The cascade process

The processes of liquefying a gas by the application of pressure and that of cooling the liquid by evaporation can be combined to form a single method of liquefying low-boiling-point gases. This is the *cascade process,* introduced in 1877 by Pictet, who used it to liquefy oxygen.

The process starts with a substance A that can be liquefied at room temperature T_R by compression alone. By pumping away the vapour under near-isentropic conditions the temperature of the liquid A can be reduced along the vaporisation curve (see Figure 16.8), the limiting temperature

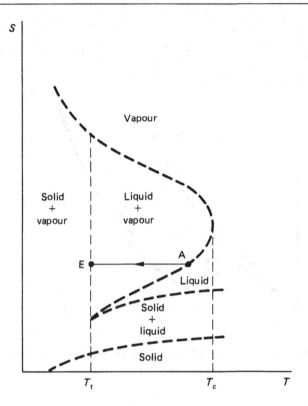

Figure 16.7 *The reduction of temperature produced by liquid evaporation under adiabatic conditions shown on the graph of the entropy S against temperature T*

obtainable being close to the triple-point temperature T_{tA}. Another substance B, whose critical temperature T_{cB} is below room temperature [1] but above T_{tA} may be cooled to T_{tA} and then liquefied at the same temperature by compression. Subsequent evaporation of liquid B will reduce its temperature to a value close to the triple point temperature T_{tB}. Values of T_c for some common gases are given in Table 16.1.

The cascade method was used extensively by Kamerlingh Onnes, who, in the period 1892–1894, developed a three-stage process that resulted in the liquefaction of oxygen, the intermediate stages using methyl chloride and ethylene. In each stage the gas circulated in a closed cycle, the vapour pumped off being recompressed and then liquefied again.

Keesom introduced a four-stage cycle, using ammonia, ethylene and methane [2], that enabled atmospheric air to be liquefied, but this is the limit of the technique: there are no liquefied gases that boil below the critical temperatures of neon, hydrogen and helium.

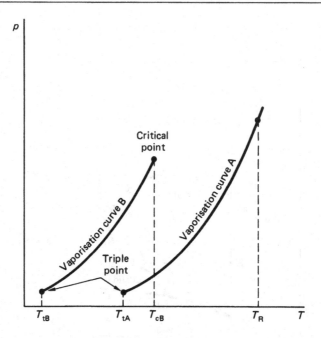

Figure 16.8 *Illustrating the cascade process for producing low temperatures*

16.4.3 The Linde and Hampson process

In 1895 Linde in Germany, Hampson in England and Tripler in the USA realised, independently, that the Joule–Thomson process could be used to liquefy gases. When the initial temperature of a gas is below the maximum inversion temperature, a fall in temperature may always be produced by an expansion of the gas through the throttle, provided that p_i and p_f are chosen correctly, but this temperature drop is always very small (see the example at the end of the chapter) and quite insufficient to liquefy the gas. However, the Joule–Thomson process involves a continuous circulation of the gas and the fall in temperature can be made cumulative, so that the temperature of the gas approaching the throttle is progressively lowered and liquefaction eventually occurs. Using this approach, hydrogen was liquefied by Dewar in 1898 and helium by Kamerlingh Onnes in 1908.

Linde's liquefier is shown schematically in Figure 16.9(a). The gas to be liquefied is compressed in two stages, first to a pressure of 20 atm and then to 200 atm. These compressions raise the temperature of the gas, which is brought back to its original value by passing the gas through the cooler. In a model of the Linde process these two stages would be treated as reversible and isothermal and represented by ab and bc on the entropy S

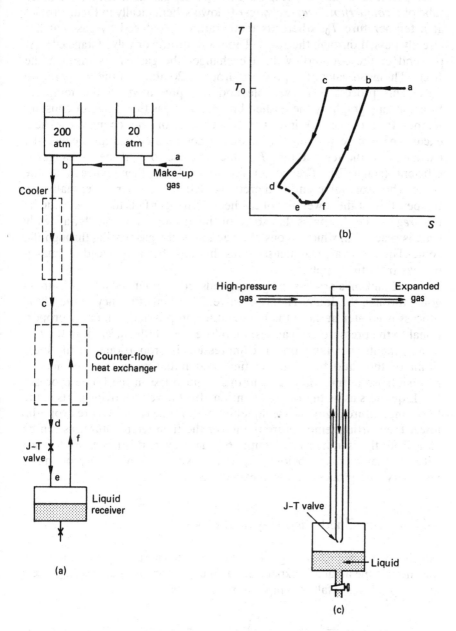

Figure 16.9 *(a) A schematic representation of the Linde–Hampson liquefier; (b) the temperature T – entropy S chart for the Linde–Hampson process; (c) a schematic representation of a counter-flow heat exchanger*

– temperature T diagram (Figure 16.9b). The gas then enters the inner tube of a *counterflow heat exchanger* (shown schematically in Figure 16.9c) at a temperature T_0, where its temperature is reduced by gas that has already passed through the Joule–Thomson throttle or valve (stage cd). At the end of the counterflow heat exchanger the gas passes through the Joule–Thomson valve, the pressure drops to 20 atm and the temperature falls. This process is irreversible and is represented on the temperature–entropy graph by the dashed line de. The gas then flows at constant temperature (stage ef) into the outer tube of the counterflow heat exchanger, where it cools the incoming compressed gas and leaves the exchanger at the temperature T_f, which is close to T_0 if the exchanger is efficient (stage fb). The cooler compressed gas then expands at the Joule–Thomson valve and is thereby cooled still further. Eventually the temperature at the lower end of the heat exchanger falls to such a value by this 'regenerative' process that some of the gas liquefies. Finally, a steady state is reached in which a constant fraction of the gas passing through the Joule–Thomson value condenses. Gas that liquefies is replaced by 'make-up' gas from the supply.

In Hampson's process the gas expands from about 200 atm to 1 atm in going through the Joule–Thomson valve. The total efficiency of the Linde process is greater because the temperature drop is approximately proportional to the pressure drop and is, therefore, very little different in the two arrangements, but the work of compression is proportional to the logarithm of the ratio between the final and initial pressures, assuming a reversible isothermal process, and this is much less in the Linde process.

Liquefiers using the Joule–Thomson effect have no moving parts in the low-temperature region of the liquefier, and, for a given pressure drop, the lower the starting temperature the lower the final temperature, as can be seen from the isenthalps of Figure 16.6. However, it is necessary to start with the temperature below T_{imax}, and when extensive precooling is necessary, the process becomes expensive.

16.4.4 The Claude and Heylandt process

An alternative approach to gas liquefaction is to cause the gas to undergo a quasireversible adiabatic expansion in an engine or turbine. Such a process always produces a fall in temperature, given by

$$\left(\frac{\partial T}{\partial V}\right)_S = -\left(\frac{\partial T}{\partial S}\right)_V\left(\frac{\partial S}{\partial V}\right)_T = -\frac{T}{C_V}\left(\frac{\partial p}{\partial T}\right)_V \qquad (16.31)$$

In the arrangement introduced by Claude in 1902 for the liquefaction of air, the air is first compressed to about 40 atm and passed through a cooler. Part of the air is then diverted to a lagged engine, where it undergoes an expansion and its temperature falls. The remainder of the air passes through a heat exchanger into a condenser, where, still at a high pressure, it is cooled by the expanded gas and readily liquefies. In this process the temperature fall per unit pressure drop in the engine decreases as the temperature decreases, and there are problems in lubricating the engine. Heylandt (1902) used much higher pressures ($\sim$ 200 atm) and the expansion engine was not insulated. Later forms of both the Claude and Heylandt machines incorporated a Joule–Thomson valve for the final condensation. This combined process is the basis for most modern gas liquefiers, such as that developed by Collins, which uses two expansion engines in a Claude-cycle liquefier and is capable of producing liquid helium with liquid nitrogen precooling only.

16.5 The measurement of low temperatures

The major requirement in the measurement of temperature is an instrument that is capable of realising thermodynamic temperatures directly. Such instruments are often called *primary thermometers*. Measurements are made of a convenient property of the thermometric substance which changes with temperature in a manner that is well understood physically.

Gas thermometers (see Section 9.3) are the basic primary thermometers and, as can be seen from Table 16.1, down to 5.2 K thermodynamic temperatures may be determined using a gas thermometer with ^{4}He at any arbitrarily chosen pressure as the thermometric substance. Suitable corrections for the non-ideality of the gas must, of course, be applied. At lower temperatures gas thermometry is still possible, but must be carried out with very dilute gas. A lower limit of about 1 K is placed on the range by the low pressures involved, the effect of the 'dead space' and the adsorption of gas on the walls of the bulb. In a recent design of constant–volume gas thermometer the 'dead space' and absorption problems were alleviated by using a large bulb ($\sim 10^3$ cm^3), the interior of which was gold-plated and polished. The capillary had a bore of 1 mm and terminated in a diaphragm that allowed constancy of volume and pressure of the gas to be precisely determined. The pressure was measured with a pressure balance in which the pressure of the gas was balanced by a weight acting over the area of a piston in a matched piston and cylinder assembly.

Another primary thermometric system in use at low temperatures is the *paramagnetic salt thermometer*, which uses the process of adiabatic demagnetisation (see Section 12.4). Here T is obtained from

$$T = \Delta Q/\Delta S$$

$$= \frac{dQ}{dT^*} \cdot \frac{dT^*}{dS}, \text{ evaluated with } B_a = 0$$

where the heat transfer dQ and entropy change dS are determined for a (reversible) isothermal process in a Carnot cycle and T^* is a magnetic temperature obtained by assuming that Curie's law holds in the form

$$T^* = \frac{a}{\chi_m} \text{ (see Equation (12.55)}$$

Thermodynamic temperatures may also be determined by measuring the acoustic wave velocity c_s in a gas. For an ideal gas

$$c_s = \sqrt{\left(\frac{\gamma RT}{M_m}\right)}$$

where M_m is the mass of 1 mol of the gas, R is the gas constant and γ is the ratio of the molar heat capacities, $C_{p,m}/C_{V,m}$. A correction is needed for the non-ideality of the gas.

Primary thermometers are not generally used in routine measurements, where there is a need for more convenient instruments, *secondary thermometers,* which must be calibrated. An important secondary thermometer is that using the vapour pressure of liquid helium. This is convenient because low temperatures are often produced and varied by pumping on a bath of liquid helium, and the saturation vapour pressure is a direct measure of the temperature of the bath. Useful semiempirical formulae connecting vapour pressure p and temperature T may be obtained using the result that, along the vaporisation curve,

$$\int g_l dT = \int g_g dT$$

where g_l and g_g are the specific Gibbs functions of the liquid and vapour phases, respectively. Vapour pressure thermometers are useful from the normal boiling point down to about 1 K for ^{4}He and about 0.5 K for ^{3}He. The main problems are that a long time is needed to achieve equilibrium and that the vapour pressure is sensitive to the presence of impurities.

Certain resistance thermometers are useful secondary thermometers in the low-temperature range. Rhodium–iron provides an instrument that is useful down to about 0.5 K and is very stable. Thermometers based on the resistance of germanium are highly sensitive and have good reproducibility, and the useful range can be varied by doping. These thermometers are useful down to about 0.3 K, but they show high magnetoresistance and it is difficult to fit an equation to the characteristic. Carbon resistors are also used as secondary thermometers down to about 0.01 K. They are reputed to be less stable than germanium resistors, but are not so sensitive to magnetic fields.

■ EXAMPLE

Q. For moderate and low pressures the equation of state of argon at a pressure p and temperature T may be written

$$pV_m = RT + Bp$$

where V_m is the molar volume, R is the gas constant and B is a function of temperature only. At s.t.p. the value of B is $-22 \times 10^{-6}\,m^3\,mol^{-1}$ and of dB/dT is $0.25 \times 10^{-6}\,m^3\,mol^{-1}\,K^{-1}$. Obtain a value for the Joule–Thomson coefficient of argon at s.t.p. if the molar heat capacity at constant pressure is a constant equal to $5\,R/2$.

A. The Joule–Thomson coefficient μ is given by Equation (16.16), which, for 1 mol of molecules of gas, may be written

$$\mu = \frac{1}{C_{p,m}}\left[T\left(\frac{\partial V}{\partial T}\right)_p - V_m \right]$$

For the given equation of state

$$\left(\frac{\partial V}{\partial T}\right)_p = \frac{R}{p} + \frac{dB}{dT}$$

so that

$$\mu = \frac{1}{C_{p,m}}\left[T\frac{R}{p} + T\frac{dB}{dT} - \frac{RT}{p} - B \right]$$

$$= \frac{1}{C_{p,m}}\left[T\frac{dB}{dT} - B \right]$$

At s.t.p. $T = 273$ K, so that, substituting the given values for $C_{p,m}$, B and dB/dT and putting $R = 8.314\,J\,K^{-1}\,mol^{-1}$, gives

$$\mu(\text{s.t.p.}) = \frac{2}{5 \times 8.314}[273 \times 0.25 \times 10^{-6} - (-22 \times 10^{-6})]$$

$$= 4.3 \times 10^{-6}\,K\,Pa^{-1}$$

or

$$\mu(\text{s.t.p.}) \approx 0.43\,K\,atm^{-1}$$

■ EXERCISES

1   Derive an expression for the Joule coefficient of a van der Waals gas.

2   When a gas undergoes a throttling process from a state characterised by
    pressure p_i and temperature T_i to one with coordinates p_f and T_f, the value
    of T_i is fixed by the precooling liquid used and that of p_f by the design of
    the apparatus. Show that T_f has the lowest value when the value of p_i
    satisfies the condition

    $$\left(\frac{\partial T}{\partial p}\right)_{H; p = p_i} = 0$$

3   Show that the Joule–Thomson coefficient μ may be expressed in the form

    $$\mu = \frac{1}{C_p}\left[-\left(\frac{\partial U}{\partial p}\right)_T - \left(\frac{\partial (pV)}{\partial p}\right)_T\right]$$

    where U is the internal energy of the gas when its pressure is p, its
    temperature T and its volume V, and C_p is the principal heat capacity at
    constant pressure. Comment on the form of this result.

4   Show that, for a van der Waals gas, the maximum inversion temperature
    T_{imax}, the Boyle temperature T_B and the critical temperature T_c are related
    by the equations

    $$T_{imax} = 2T_B = \frac{27}{4} T_c$$

5   A certain gas has a molar heat capacity at constant pressure of $5R/2$,
    independent of temperature, and an equation of state

    $$p(V_m - b) = RT$$

    where V_m is the molar volume at a temperature T when the pressure is p,
    and b is a constant equal to 25×10^{-6} m^3.
        Calculate the change in temperature of the gas when it undergoes a
    Joule–Thomson expansion from a pressure of 50×10^5 Pa to one of 2×10^5
    Pa.

6   Use the reduced equation of van der Waals (Equation 13.27) to show that
    the equation of the inversion curve for a van der Waals gas is

    $$\hat{p} = \frac{9(2\hat{V}_m - 1)}{\hat{V}_m^2}$$

    where $\hat{p}$ and $\hat{V}_m$ are, respectively, the reduced pressure and reduced molar
    volume.

Notes

1 T_c may be estimated using van der Waals' equation: the parameters a and b in that equation may be obtained from the observed isotherms for $T > T_c$ and T_c may be obtained from $T_c = 8a/27$ Rb.

2 Table 16.1 shows that the methane stage is not strictly necessary, but its introduction improved the thermodynamic efficiency of the four-stage process as a whole by making it a little closer to a reversible process.

17

A simple kinetic theory of gases

One of the important features of classical thermodynamics is that it has very few independent quantities that need to be 'explained' in terms of the structure of matter. The thermodynamically reversible behaviour of all systems with two degrees of freedom can be described in terms of the equation of state and one principal heat capacity. For a closed hydrostatic system the principal heat capacities are C_p and C_V, the heat capacities at constant pressure and constant volume, respectively, and the equation of state is a relation between pressure p, volume V and temperature T for equilibrium states. Therefore, for a microscopic description of a closed hydrostatic system it is only necessary to describe in terms of the structure of matter either C_p or C_V and any two of p, V and T. All other equilibrium properties of the system may be derived from these by thermodynamic reasoning.

The kinetic theory of matter seeks, as a principal aim, to show how the properties of matter in bulk may be explained in terms of the interactions of the constituent particles, and the success of this approach will depend on attributing suitable properties to these constituent particles. In this chapter this aim will be examined for gases and, especially, for gases at low pressure, when their macroscopic behaviour is particularly simple.

17.1 A model of a gas

The idea that matter is made of small discrete particles, called atoms, goes back to the Greeks, but the modern concept of the atom was suggested by the simple regularities that are observed in chemical experiments and are contained in the law of constant proportions and the law of multiple proportions. The law of constant proportions states that a definite chemical compound always contains the same elements in the same proportions by mass, while the law of multiple proportion states that when two elements combine in more than one proportion, there is a simple relationship between the masses of the other element present. For gases simple laws hold not only for the masses of the reacting substances, but also for their volumes. This result is contained in Gay-Lussac's law (1809), which states that gases combine chemically in simple ratios by volume and that the volume of the gaseous product bears a simple ratio to the volumes of the reacting gases.

Dalton (1808) realised that these laws receive a ready explanation in terms of an atomic model of matter. His model assumed that matter is composed of very small discrete particles, which he called atoms, which retain a separate identity in chemical processes. Atoms of a given chemical element are all identical in size and mass and are able to combine with atoms of other elements in simple numerical ratios. The concept of the atom was extended by Avogadro (1811), who introduced the important distinction between atoms and aggregates of atoms, which he called molecules. A molecule may be defined as the smallest entity able to have a stable existence in the gaseous phase. For convenience, the term 'molecule' will also be used here to describe the constituent particles of matter in the solid and liquid phases, even when these are atoms or ions (charged atoms).

On the kinetic model a solid of macroscopic dimensions is assumed to consist of a very large number of molecules vibrating about fixed positions. When these fixed points form a regular three-dimensional array, the solid is termed crystalline.

Attempts to stretch solids are opposed by attractive forces between the molecules, while attempts to compress solids bring repulsive forces into play. These repulsive forces increase rapidly with decreasing molecular separation (solids are not easily compressed), which suggests that, in solids, the molecules, pictured as slightly soft spheres, are in contact with their nearest neighbours.

As the temperature of the solid is increased, the average amplitude of the molecular vibration increases and molecules are able to break away from their fixed positions. This is the process of melting, which, for a pure substance, occurs at a fixed temperature for a given pressure. The density of a liquid is only slightly less than that of the parent solid: the molecules are still in contact, giving the liquid a definite volume at a given pressure, but the number of nearest neighbours is slightly reduced and the molecules are not fixed in position. Fluctuations in the energy of individual molecules enable some molecules to escape from the liquid surface. This is evaporation, which becomes more pronounced as the temperature of the liquid rises.

Eventually a temperature is reached at which all the molecules leave the liquid, to form a gas. At s.t.p. the gas density is about 1/1000 of the density of the solid under the same conditions, indicating that the molecules in a gas are, on average, very much further apart than they are in the condensed phases. The molecules then exert negligible forces on each other, except during collisions. If it is assumed that, on average, the molecules travel with high speeds, they will travel in straight lines between collisions, so that a gas is able to fill all the space available to it.

17.2 The pressure exerted by a gas

A simple model of a one-component gas assumes that the gas is composed of molecules; that all the molecules are identical, having the same mass; and that they move with continuous random motion, so that the molecules are uniformly distributed in the volume that they occupy and the velocity distribution of the molecules is isotropic.

The earliest model for the pressure exerted by a gas (Gassendi, mid-seventeenth century; D. Bernoulli, 1738) pictured the pressure as arising from the large number of impacts per second of the tiny molecules with the walls of the container. This is still the basis of the current view of how a pressure is exerted by a gas, but to develop a simple model some further aspects of molecular behaviour must be examined.

A gas may be condensed to the liquid phase by lowering the temperature below the critical temperature and increasing the pressure appropriately, indicating that attractive (cohesive) forces exist between the molecules when they are sufficiently close together. On the other hand, liquids are not easily compressed, indicating that strong repulsive forces are set up when molecules are brought very close. For two molecules of any substance the force of interaction F varies with the separation d between the centres in the general way shown in Figure 17.1, where repulsive forces are counted positive. When the separation is greater than d_1, the attractive force between the molecules is negligible, and at the equilibrium separation d_0 the net force is zero. For separations less than d_0 the mutual force is repulsive and increases rapidly with decrease in d. When the average molecular separation is large, as in a dilute gas, the attractive force may be neglected. Further, if the molecules are treated as hard spheres, the repulsive force may be approximated by the dashed line shown in Figure 17.1, so that repulsive forces are only brought into play during molecular collisions — that is, when the separation of the molecules equals the molecular diameter σ. For a dilute gas the volume occupied by the molecules is a very small fraction of the total volume occupied by the gas.

To investigate the collisions of the gas molecules with the walls of the container, Knudsen (1915) allowed a narrow beam of molecules to be 'reflected' from a small area of the inner surface of a hollow, evacuated sphere, as shown in Figure 17.2. This small area was maintained at room temperature while the remainder of the surface was kept at a low temperature, so that the molecules remained where they struck. From his observations Knudsen concluded that, for 'normal' surfaces, the direction of molecular motion after impact is not related to that before impact.

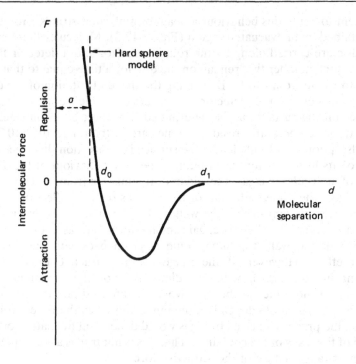

Figure 17.1   *The force F between two molecules as a function of their separation* d

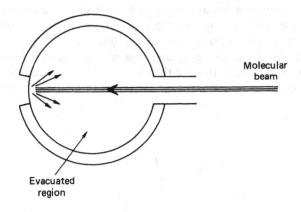

Figure 17.2   *Apparatus to investigate the collision of gas molecules with the walls of the container*

Rather, following impact, moelcules start out in random directions. This conclusion is consistent with the results of experiments showing that, at all temperatures, a gas molecule striking a solid surface is adsorbed to the surface and resides there for a certain time before leaving. In a simple experiment to study this behaviour a beam of molecules strikes a rotating disc contained in an evacuated region (Figure 17.3). Molecules that 'stick' to this disc are carried along by the rotating surface and a detector then determines the number that remain on the disc for a time equal to that for the disc to rotate from I to D. By varying the speed of rotation of the disc and the position of the detector, the average residence time of the molecules on the surface can be determined. For inert gases on steel or glass at room temperature, residence times are in the range 10^{-6}–10^{-9} s. After adsorption a molecule leaves the surface in a direction that does not depend on its history before adsorption. When the behaviour of the large number of molecules in a gas is considered, re-emission from the container wall produces the same molecular distribution as regular reflection.

As a molecule approaches the wall of the container, it experiences a force of attraction and, therefore, gains momentum, but, as a result of the collision with the wall, this momentum is given back and there is no resultant effect. However, if the range of the attractive forces were significant in comparison with the dimensions of the container, the momentum of molecules in the bulk would be affected and the pressure exerted by the gas would depend on the range and strength of these forces — that is, the pressure exerted by a gas would depend on the nature of the material of the walls of the container. That this is not true is a consequence of the short-range nature of the attractive forces.

When the temperature of a gas is constant, the mean speed of the molecules must remain constant. Therefore, collisions between gas molecules and also between gas molecules and the wall of the container must, on average, be elastic — that is, the effects of inelastic and superelastic collisions, with respect to translational motion, must cancel out.

On the basis of the above assumptions an expression for the pressure exerted by a gas may be derived in the following way. Let the gas consist of identical molecules, each of mass m, and let there be n molecules per unit

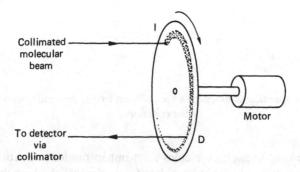

Figure 17.3 *Apparatus to determine the residence time of a gas molecule on a surface*

volume. These molecules may be divided into groups according to their velocity. Let group 1 consist of those molecules that have velocity components u_1, v_1, w_1 in the x-, y-, z-directions, respectively. Group 2 will consist of those molecules with velocity components u_2, v_2, w_2, and so on. Let the number of molecules per unit volume in group 1 be n_1.

Consider now the molecules striking a small area dA of the wall of the container, assumed to be normal to the x-direction. The molecules in group 1 that strike this area dA in a time dt are contained in a cylinder of base area dA, having prismatic surfaces parallel to the direction of the resultant velocity with components u_1, v_1, w_1, and of height $u_1 dt$ (see Figure 17.4). This cylinder has a volume $u_1 dt dA$ and contains $n_1 u_1 dt dA$ molecules from group 1. Since the distribution of molecular velocities as a whole is random, at any instant half the molecules with velocity components u_1, v_1, w_1 will be moving towards dA and half will be moving away.

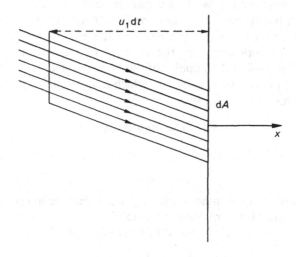

Figure 17.4 *To calculate the pressure exerted by a model gas.*

The momentum brought up to the surface dA in a time dt by the molecules of group 1 is, therefore, $\frac{1}{2} mu_1 \cdot n_1 u_1 dt dA$, which is $\frac{1}{2} n_1 m u_1^2 dt dA$. The total momentum in the direction perpendicular to dA brought up to the element in a time dt is obtained by summing the expression for group 1 over all groups of molecules for which the value of u is positive. Therefore, the total momentum in the positive x-direction brought up to the element of area dA in a time dt is

$$\frac{1}{2} m dt dA \sum_i n_i u_i^2$$

which can immediately be written

$$\tfrac{1}{2} m dt dA n \bar{u}^2$$

where

$$\bar{u}^2 = \frac{n_1 u_1^2 + n_2 u_2^2 + n_3 u_3^2 + \cdots}{n}$$

Therefore, the total momentum perpendicular to dA brought up to it in a time dt may be written

$$\frac{1}{2} m n \bar{u}^2 dt dA$$

When conditions are steady, there is no net gas flow in any direction and the total momentum in the x-direction must be exactly reversed by collisions with the element of surface, whatever the precise details of the process. The total change in momentum in the x-direction which occurs in a time dt for collisions with the element of area dA is, therefore, $mn\bar{u}^2 \, dt dA$. This equals the impulse of the force exerted by the molecules on the element of area dA in a time dt, which is also equal to $pdAdt$, where p is the gas pressure. Therefore,

$$pdAdt = mn\bar{u}^2 dt dA$$

or

$$p = mn\bar{u}^2 \tag{17.1}$$

It is not possible to determine $\bar{u}^2$ directly, but it may be inferred from the *mean square speed* if the molecular motion is random.

Let $\bar{c}^2$ be the mean square speed of the molecules, defined by

$$\bar{c}^2 = \frac{n_1 c_1^2 + n_2 c_2^2 + n_3 c_3^2 + \cdots}{n_1 + n_2 + n_3 + \cdots}$$

where c_1 is the speed of the molecules in group 1.

Expressing $\bar{c}^2$ in terms of the velocity components gives

$$\bar{c}^2 = \frac{n_1 u_1^2 + n_1 v_1^2 + n_1 w_1^2 + n_2 u_2^2 + n_2 v_2^2 + n_2 w_2^2 + \cdots}{n}$$

$$= \frac{n_1 u_1^2 + n_2 u_2^2 + \cdots}{n} + \frac{n_1 v_1^2 + n_2 v_2^2 + \cdots}{n} + \frac{n_1 w_1^2 + n_2 w_2^2 + \cdots}{n}$$

or

$$\bar{c}^2 = \bar{u}^2 + \bar{u}^2 + \bar{w}^2$$

Since the motion of the molecules is assumed to be completely random,

$$\overline{u^2} = \overline{v^2} = \overline{w^2}$$

Therefore,

$$\overline{u^2} = \frac{1}{3}\overline{c^2} \tag{17.2}$$

and

$$p = \frac{1}{3}mn\overline{c^2} = \frac{1}{3}\rho\overline{c^2} \tag{17.3}$$

where ρ is the density of the gas.

Equation (17.3) allows the *root mean square speed* to be calculated for any gas for which this kinetic model is a good representation, since the pressure and density are readily measured. Table 17.1 gives the value of $(\overline{c^2})^{1/2}$ at s.t.p. for some common gases. It should be noted that the calculated speed is greater than the speed of sound under these conditions, but of the same order of magnitude.

Table 17.1 The root mean square speed $(\overline{c^2})^{1/2}$ and the speed of sound c_s, both determined at s.t.p., for some common gases

Gas	$(\overline{c^2})^{1/2}/\text{ms}^{-1}$	c_s/ms^{-1}
Hydrogen	1838	1286
Oxygen	461	332
Nitrogen	493	337
Argon	413	307.8
Helium	1307	971.9
Carbon dioxide	393	259 (low-frequency)
		268 (high-frequency)

Values of c_s from Kaye and Laby (1973).

The above discussion has 'explained' the pressure exerted by a dilute gas in terms of the behaviour of its constituent molecules. It is now necessary to introduce the concept of temperature.

Equation (17.3) may be written

$$p = \frac{1}{3}\rho\overline{c^2} = \frac{2}{3}\cdot\frac{1}{2}\rho\overline{c^2} \tag{17.4}$$

or

$$p = \frac{2}{3}(\text{kinetic energy of the molecules in unit volume}) \tag{17.4a}$$

For 1 mol of molecules of the gas, Equation (17.4) becomes

$$p = \frac{2}{3} \cdot \frac{1}{2} \frac{M_m}{V_m} \overline{c^2} \tag{17.5}$$

where M_m is the mass of 1 mol of molecules and V_m is the volume occupied when the pressure is p.

$\frac{1}{2} M_m \overline{c^2}$ is the total translational kinetic energy of all the molecules in 1 mol. Since the model used is that of a dilute gas, Equation (17.5) may be compared with the ideal gas equation

$$pV_m = RT \tag{17.6}$$

where R is the gas constant. Assuming that Equations (17.5) and (17.6) both represent the behaviour of an ideal gas, or that of a real gas at vanishingly small pressures, it follows that

$$\frac{1}{2} M_m \overline{c^2} = \frac{3}{2} RT \tag{17.7}$$

If N_A is Avogadro's constant,

$$M_m = mN_A$$

and, therefore,

$$\frac{1}{2} mN_A \overline{c^2} = \frac{3}{2} RT$$

or

$$\frac{1}{2} m\overline{c^2} = \frac{3}{2} \frac{R}{N_A} T = \frac{3}{2} kT \tag{17.8}$$

where k is Boltzmann's constant.

Therefore, the average translational kinetic energy of a molecule in an ideal gas (dilute real gas) at a temperature T is $\frac{3}{2}kT$, which is proportional to T. This result provides a molecular interpretation for the temperature of dilute gases: the thermodynamic temperature of a dilute gas (ideal gas) is proportional to the mean kinetic energy of translation of its molecules. It should be noted from Equation (17.8) that the mean kinetic energy of translation of the molecules of a gas does not depend on the molecular mass. If two gases of different molecular mass, but having the same temperature, are mixed, there is no change in the average kinetic energy of translation of the molecules and, therefore, no change in temperature.

Consider now in more detail the flux of molecules striking the container wall. In a time dt the molecules that strike a surface element of

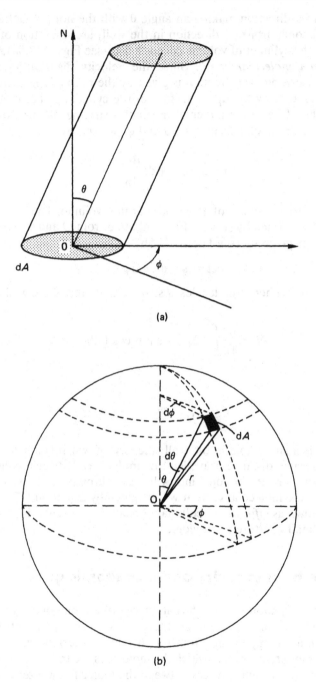

(a)

(b)

Figure 17.5 *(a) To determine the flux of molecules striking the wall of the container; (b) the elementary solid angle $d\Omega = dA/r^2$*

area dA from a direction making an angle θ with the normal ON and an angle φ with some arbitrary direction in the wall, are a fraction of those contained in the cylinder of volume $\bar{c}dt \cos \theta\, dA$, (see Figure 17.5a), where $\bar{c}$ is the *mean molecular speed*. Since the velocity distribution of the molecules is isotropic, this fraction is given by the ratio of the solid angle dΩ subtended at O by the upper surface of the cylinder to the total solid angle 4π. Therefore, the number of molecules striking the area dA in a time dt from a direction specified by θ and φ is given by

$$n\bar{c} \cos \theta\, dA\, dt\, \frac{d\Omega}{4\pi}$$

where n is the total number of molecules per unit volume. The elementary solid angle dΩ, formed by straight lines radiating from O and touching the edges of dA, is given by (see Figure 17.5b)

$$d\Omega = \sin \theta\, d\theta\, d\varphi$$

Therefore, the number of molecules N striking unit area of the wall in unit time is given by

$$N = \frac{n\bar{c}}{4\pi} \int_0^{2\pi} d\varphi \int_0^{\pi/2} \sin \theta \cos \theta\, d\theta$$

or

$$N = \frac{n\bar{c}}{4} \tag{17.9}$$

If there is a small hole in the wall, the size of which is much smaller than the average distance travelled by molecules between collisions, molecules arriving at the opening will pass through without making collisions and the molecular current will be given by Equation (17.9). This process is known as *effusion*. A collimated beam of molecules is produced if a second small aperture is provided.

17.3 The heat capacity of a monatomic gas

The major contribution to the internal energy of a monatomic gas comes from the translational kinetic energy, there being a very small contribution to the potential energy arising from the attractive intermolecular forces. For an ideal monatomic gas, in which the molecules are treated as very tiny hard spheres, the attractive forces between the molecules are zero and the repulsive forces act merely to change the direction of motion, provided that rotational effects may be neglected. The translational kinetic energy of the gas molecules may then be taken as the internal energy U. Equation (17.7) shows that the total translational kinetic energy of the model gas is proportional to the temperature T — that is,

$$U = \text{constant} \cdot T$$

or

$$\left(\frac{\partial U}{\partial V}\right)_T = 0 \qquad (17.10)$$

Equation (17.10) is Joule's law, which holds for ideal gases. For 1 mol of molecules of ideal monatomic gas the internal energy is given by

$$U_m = \frac{3}{2} RT \qquad (17.11)$$

It has been shown in Section 11.2 that, for a closed hydrostatic system, the heat capacity at constant volume C_V may be written

$$C_V = \left(\frac{\partial U}{\partial T}\right)_V, \text{ so that } C_{V,m} = \left(\frac{\partial U_m}{\partial T}\right)_V$$

For an ideal gas this equation may be written

$$C_{V,m} = \frac{dU_m}{dT} \qquad (17.12)$$

Combining Equations (17.11) and (17.12) gives

$$C_{V,m} = \frac{3}{2} R \qquad (17.13)$$

for the molar heat capacity of an ideal monatomic gas. This is the value obtained experimentally for real, dilute monatomic gases. The simple kinetic theory model of a monatomic gas is able to 'explain' $C_{V,m}$ in terms of the temperature-dependence of the average kinetic energy of the molecules. For any ideal gas $C_{p,m}$ may, of course, be obtained from the equation

$$C_{p,m} = C_{V,m} + R \qquad (17.14)$$

For diatomic and polyatomic gases there are contributions to the internal energy from rotations and vibrations of the molecules. These may be treated in a straightforward manner, but that is beyond the intention of this chapter.

17.4 The Maxwell distribution law

The discussion so far has used two averages of the molecular speed — the root mean square speed and the mean speed — but has not investigated in detail the distribution of molecular speeds. For the ideal molecules assumed by the kinetic theory this distribution is given by an expression which may be obtained in the following way.

It is assumed that the distribution of molecular velocities is isotropic and that the distribution of molecular speeds for any one component of velocity is independent of that in any other component — that is, the motion of the molecules is chaotic. Further, it is assumed that the gas is so dilute that only *binary encounters* (encounters involving two molecules only) are important. On the basis of these assumptions it follows that, for each change in velocity in a binary molecular encounter, there is a restoration of the original velocities by an inverse encounter.

In any encounter linear momentum must be conserved, so that binary collisions must be coplanar. Finally, it is assumed that translational kinetic energy is conserved. This means assuming that collisions are elastic and no energy is transferred during collisions between translational kinetic energy and internal modes of storage such as rotation and excitation.

For a gas consisting of a single species let $f(c)$ be the number of molecules per unit volume having a velocity c — that is, the number of molecules per unit volume with velocities in the range c to $c + dc$ is $f(c)dc$. Consider now a binary collision in which two molecules have their velocities changed from c_1 and c_2 to c_3 and c_4, respectively (Figure 17.6a). The probability of such a collision occurring will be proportional to the numbers of molecules per unit volume having these velocities — that is, to the product $f(c_1)f(c_2)$. Therefore, the number of such collisions per unit volume per unit time may be written $af(c_1)f(c_2)$, where a is a constant. Similarly, the number of inverse collisions per unit volume per unit time (Figure 17.6b) is $a'f(c_3)f(c_4)$, where a' is also a constant. Since the gas is in equilibrium and the velocity distribution is unchanged by collisions, these two rates must be equal. Further, the collisions appear equivalent when considered in the centre of mass frame of reference, so that $a=a'$. Then

$$f(c_1)f(c_2) = f(c_3)f(c_4)$$

or

$$\ln f(c_1) + \ln f(c_2) = \ln f(c_3) + \ln f(c_4) \qquad (17.15)$$

Since it is also assumed that kinetic energy is conserved, and as $c \cdot c = |c|^2 = c^2$

$$c_1^2 + c_2^2 = c_3^2 + c_4^2 \qquad (17.16)$$

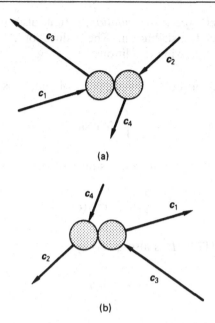

Figure 17.6 *(a) A binary molecular encounter; (b) the inverse binary encounter*

An obvious solution of Equations (17.15) and (17.16) is

$$\ln f(c) \propto c^2$$

or

$$f(c) = A \exp(-\beta c^2) \tag{17.17}$$

where A and β are constants. The minus sign is necessary to ensure that no molecules have an infinite energy.

Equation (17.17) gives the *Maxwell velocity distribution* for the molecules of the model considered. Frequently, however, it is more important to know the fraction of the molecules of a gas that move with a particular speed, irrespective of direction. Let $N(c)dc$ be the number of molecules per unit volume whose speeds lie in the range c to $c + dc$, irrespective of direction. The velocity distribution has spherical symmetry with respect to the three spatial directions, so that $N(c)\,dc$ is equal to the number of velocity vectors whose tips end in the volume shell lying between radii c and $c + dc$ — that is,

$$N(c)\,dc = 4\pi c^2 f(c)\,dc$$

or, substituting from Equation (17.17),

$$N(c) = 4\pi c^2 A \exp(-\beta c^2) \tag{17.18}$$

This is the *Maxwell speed distribution,* often also called the Maxwell–Boltzmann speed distribution. The values of A and β may be calculated using the following conditions:

1 The total number of molecules per unit volume, n, is given by

$$n = \int_0^\infty N(c)\,dc \qquad (17.19)$$

2 The total energy E of the molecules in unit volume is given by

$$E = \frac{1}{2}m\int_0^\infty c^2 N(c)\,dc \qquad (17.20)$$

3 From Equation (17.8) E is also given by

$$E = \frac{3}{2}nkT \qquad (17.21)$$

Using the standard integrals

$$\int_0^\infty x^2 e^{-ax^2}\,dx = \frac{1}{4a}\left(\frac{\pi}{a}\right)^{1/2}$$

and

$$\int_0^\infty x^4 e^{-ax^2}\,dx = \frac{3}{8a^2}\left(\frac{\pi}{a}\right)^{1/2}$$

where a is a constant, Equations (17.18) – (17.21) may be combined to give

$$N(c) = 4\pi c^2 n\left(\frac{m}{2\pi kT}\right)^{3/2}\exp(-mc^2/2kT) \qquad (17.22)$$

The dependence of $N(c)$ on c, as given by Equation (17.22), is shown in Figure 17.7 for oxygen at 273 K.

Direct investigations of the dependence of $N(c)$ on c have used a gas of metal molecules (actually atoms) produced by an oven. Molecules emerge from a small hole in the wall of the oven, and slits are used to form a collimated beam which is then analysed by some form of velocity filter. The whole apparatus is in an evacuated enclosure. In the technique used by

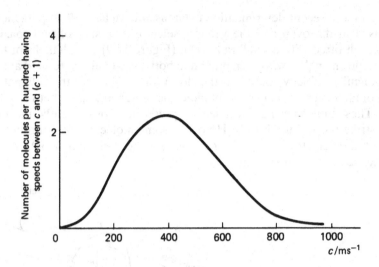

Figure 17.7 *The Maxwell–Boltzmann speed distribution for oxygen at 273 K. Taken from page 10 (Figure 3) of* An Outline of Atomic Physics, *by O. H. Blackwood, E. Hutchisson, T. H. Osgood, A. E. Ruark, W. N. St. Peter, G. A. Scott and A. G. Worthing, John Wiley and Sons, New York, second edition, 1944*

Zartman (1931) and Ko (1934) the gas was produced by raising the temperature of a bismuth sample to 800 °C and the velocity filter was a rotating drum with a slit, through which a pulse of gas molecules entered at each rotation (Figure 17.8). Fast molecules in the pulse arrive near A on the glass plate AB, on which the molecules condense. Slower molecules do not arrive at the plate until a region B has moved round onto the axis of the collimated beam of molecules. The darkening of the glass is, therefore, related to the numbers of molecules arriving and their position along the glass plate to their speeds.

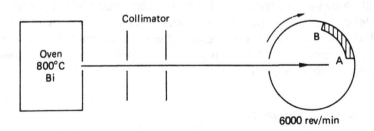

Figure 17.8 *The apparatus of Zartman and Ko for measuring the speeds of gas molecules*

In a more recent determination (Marcus and McFee, 1959) potassium was used in the oven and the velocity selector consisted of two coaxial discs, each fitted with a small radial slot (Figure 17.9). The lengths of the slots made an angle θ with each other and both discs could be rotated at the same angular velocity, ω. When the slot in disc A crosses the collimated beam of molecules, a small pulse of molecules enters the space between the discs. These molecules have a range of speeds and, consequently, most of them strike the second disc, B. However, some molecules pass through B to reach the detector. These molecules have a speed given by $L\omega/\theta$, where L is the separation of the discs.

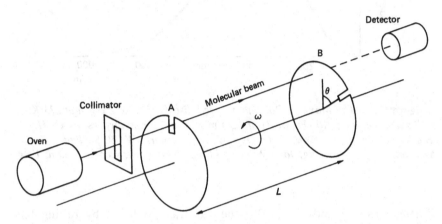

Figure 17.9   *The apparatus of Marcus and McFee for measuring the speeds of gas molecules*

It should be noted that, in direct methods of the type described, the hole in the walls of the oven from which the molecules emerge also acts as a kind of velocity filter. The number of gas molecules emerging per unit area of the hole per unit time with speeds in the range c to $c + dc$ is given by $N(c)c/4$ (see Equation 17.9), whereas the number of molecules per unit volume in the oven with speeds in the range c to $c + dc$ is $N(c)$, the Maxwell–Boltzmann distribution. Therefore, the speed distribution of the molecules emerging from the oven is skewed towards higher speeds. When this effect is taken into account, good agreement is found between the Maxwell–Boltzmann speed distribution and the speed distribution in gases composed of simple molecules.

17.5 Mean free path

Because the average separation of the molecules in a gas at moderate and low pressures is much larger than the range of the attractive intermolecular force, for most of the time a gas molecule is unaffected by the presence of the other molecules. However, when the separation of two molecules approaches σ (see Figure 17.1), the intermolecular forces cause significant deflections of the molecules and a collision is said to occur.

On the hard-sphere model of a gas molecule two gas molecules do not interact until the distance between the centres reaches σ, and then an impulsive force produces deflections. The distance that a molecule travels between collisions is known as the *free path*, and, for any particular molecule, this free path is subject to a large statistical variation from collision to collision. This distribution of free paths may be deduced in the following way.

Since, statistically, the environment of any particular gas molecule is independent of time, the probability that a collision will occur when a molecule travels a very small distance dx is proportional to dx and may be written $a dx$, where a is a constant. Then the probability that there will not be a collision when the molecule travels a distance dx is $(1 - a dx)$. Let the probability that a molecule will describe a path of at least a finite length x before making a collision be $f(x)$. Then the probability that a molecule travels a distance x and then a further distance dx before making a collision is $f(x)(1-a dx)$. But, from the definition of $f(x)$, the probability that a molecule travels a distance $x + dx$ without making a collision is $f(x + dx)$. Therefore,

$$f(x + dx) = f(x)\,(1 - a\,dx) \tag{17.23}$$

Writing $f(x + dx)$ as

$$f(x) + \frac{df(x)}{dx}dx$$

Equation (17.23) becomes

$$\frac{df(x)}{dx} = -a\,dx \tag{17.24}$$

and integrating gives

$$\ln f(x) = -ax + \text{constant}$$

or

$$f(x) = A e^{-ax} \tag{17.25}$$

where A is a constant. When x is zero, $f(x)$ must be unity, so that A is unity and

$$f(x) = e^{-ax} \tag{17.26}$$

Therefore, the probability of a free path having a length between x and $x + dx$ is $ae^{-ax} dx$.

The average value of the free path is known as the *mean free path l*, and is the average value of x weighted according to the numbers of free paths of different lengths. Therefore,

$$l = \frac{\int_0^\infty xaf(x)\,dx}{\int_0^\infty af(x)\,dx} = \frac{1}{a} \tag{17.27}$$

From Equations (17.6) and (17.27) it follows that $f(x)$ is equal to $e^{-x/l}$.

A more direct interpretation of l may be made using the hard-sphere model of a gas. To simplify the problem further, assume that the molecule under consideration is moving much more rapidly than its immediate neighbours, which may be assumed to be at rest. This is the *fast-molecule approximation*. Let the speed of the moving molecule be c. A collision occurs when the centre of a stationary molecule is at a distance equal to, or less than, the molecular diameter σ from the trajectory followed by the centre of the moving molecule A (see Figure 17.10). In unit time the moving molecule travels a distance c and, in so doing, collides with all molecules whose centres lie within a volume $\pi\sigma^2 c$. On average, there will

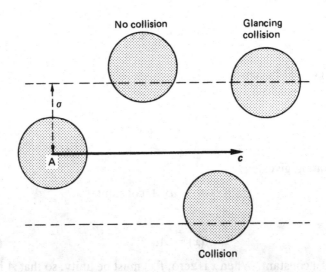

Figure 17.10 *In unit time molecule A will collide with all molecules whose centres lie within a cylinder of radius σ and length c.*

be no $\pi\sigma^2 c n$ molecules with centres in this volume. Therefore, the mean free path l is given by

$$l = \frac{c}{\pi\sigma^2 c\, n} = \frac{1}{\pi\sigma^2 n} \tag{17.28}$$

Despite the crudeness of this model, the result it gives is not much altered by assuming that the molecules move with speeds distributed according to the Maxwell–Boltzmann law. Making this assumption gives

$$l = \frac{1}{\sqrt{2}} \cdot \frac{1}{\pi\sigma^2 n} \tag{17.29}$$

Since, for a gas at constant temperature, the pressure p is proportional to n

$$l = \frac{\text{constant}}{p} \tag{17.30}$$

Values of l may be obtained from measurements of the gas viscosity [1] and, for common gases at s.t.p., l is of the order of 10^{-7} m. The mean time between collisions for a molecule is given by $l/\bar{c}$, where $\bar{c}$ is the mean speed, and for air at s.t.p. this interval is about 100 ps.

When l and n are known, the value of σ may be obtained from Equation (17.29). The value obtained is known as the *gas-kinetic molecular diameter* and is of the order of 10^{-10} m for common gases.

■ EXAMPLE

Q. Determine an expression for the mean free path of a gas molecule that is moving very slowly compared with its immediate neighbours. This is the *slow-molecule approximation*.

A. Let the speed of the molecule under consideration be c'. This molecule may be treated as a stationary target that is being bombarded by the neighbouring molecules. Assuming a hard sphere model for the gas molecules, a collision will occur when the intermolecular separation is equal to the gas-kinetic diameter σ. Treating the bombarding molecules as point-masses, the effective surface on which collisions occur is a sphere, centred on the stationary molecule S and of surface area $4\pi\sigma^2$ (see Figure 17.11). From Equation (17.9), the number of bombarding molecules striking this area in unit time is $1/4\ (\bar{c}4\pi\sigma^2)$, where $\bar{c}$ is the mean molecular speed. Therefore, the mean time between collisions, $\bar{t}$, is given by

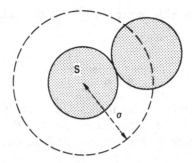

Figure 17.11 *Collisions with the stationary molecule S*

$$\bar{t} = \frac{1}{nc\pi\sigma^2}$$

and in this time the nearly stationary molecule travels a distance equal to its mean free path l' — that is,

$$l' = \frac{c'}{nc\pi\sigma^2}$$

■ EXERCISES

1 Determine the number of molecules in 1 m³ of an ideal gas at s.t.p. — that is, at a temperature of 273 K and a pressure of 1 atm (1.01×10^5 Pa).

   Boltzmann's constant = 1.381×10^{-23} J K^{-1}.

2 Calculate the mean free path of a molecule in a sample of nitrogen gas at s.t.p. The molecular diameter is 3.0×10^{-10} m.

3 For all gases the molar volume at s.t.p. is 22.4×10^3 cm³. The relative molecular mass of nitrogen is 28. Using the result from Exercise 1, calculate the root mean square molecular speed at s.t.p. The gas constant is 8.31 J K^{-1} mol^{-1}.

4 Distinguish between the mean free path and the mean molecular separation in a gas. Show that

$$\frac{\text{mean free path}}{\text{mean molecular separation}} \approx \frac{1}{\pi\sigma^2 n^{2/3}}$$

   where n is the number of molecules per unit volume and σ is the gas-kinetic diameter.

Note

1. See, for example, *Kinetic Theory* by J. M. Pendlebury (Bristol: Adam Hilger, 1985).

18

Heat transfer

In classical thermodynamics the term 'heat' is used to describe the energy transfer that occurs between a system and its surroundings when a temperature difference exists between them. It follows that heat transfer can only take place across the boundary of a system. Thermodynamics is able to describe the direction of the heat transfer across the boundary and the total mount of energy transferred in this way, but not the rate at which the transfer takes place.

When a system is in a non-equilibrium state, there may be a temperature gradient within it. It such a situation there is a transfer of energy from regions of high temperature to regions of lower temperature. Strictly, this transfer is a transfer of internal energy, but in common usage it is referred to as a flow of heat and this usage will be retained for this chapter, where the interest is in the rate at which the energy transfer takes place.

Three processes are recognised by which heat transfer takes place in a system — namely, *conduction, convection* and *radiation.* Conduction is that process which transports heat in a medium from one location to another without any relative bulk movement of parts of the medium. This energy transport is, of course, from regions of high temperature to regions of lower temperature. Conduction is usually the only heat transfer process in solids, although it also occurs in liquids and gases.

In liquids and gases, however, energy transfer may also occur by parts of the fluid moving relative to others and carrying energy along. This is the process known as convection.

All bodies emit electromagnetic radiation by virtue of their temperature. This is known as thermal radiation. Radiation processes are able to transfer net amounts of energy between bodies at different temperatures, so that radiation constitutes another form of heat transfer.

These three processes will now be considered in turn, but it should be realised that in many situations they operate simultaneously and each affects the total rate of energy transport. In addition, evaporation and condensation may occur, and in many circumstances they dominate the energy transfer.

18.1 Thermal conductivity

From a macroscopic standpoint, the conduction of heat through a material is described by the rate equation introduced by Fourier in 1822. Consider a small plane surface of area A drawn within a medium. The rate of conduction of heat dq/dt through the area is given by

$$\frac{dq}{dt} = - kA\frac{\partial T}{\partial n} \tag{18.1}$$

where $\partial T/\partial n$ is the temperature gradient measured normal to the area, and the negative sign indicates that the direction of heat flow is down the temperature gradient. Equation (18.1) is the Fourier definition of a quantity k, known as the *thermal conductivity* of the material, whose properties must be determined by experiment. The units of k are $J\,s^{-1}\,m^{-1}\,K^{-1}$ or $W\,m^{-1}\,K^{-1}$. It will be assumed that the material is isotropic, so that k is a scalar.

Experiment shows that k is independent of $\partial T/\partial n$ and, except at very low temperatures, is also independent of A. Further, k has the same value for both steady-state and changing conditions and is also a function of temperature.

18.2 General results

The variation of k with temperature is shown for a number of materials in Figure 18.1. For solid metallic elements k changes only slowly over a wide range of temperature, except at low temperatures, where k first rises and then falls towards zero as the temperature approaches absolute zero. Apart from this low-temperature behaviour, solid metals are, in general, good thermal conductors.

Gases, solid alloys and glasses have values of k which increase with increasing temperature. Gases are, in general, very poor thermal conductors.

Non-metallic liquids are moderate thermal conductors and k usually increases as the temperature is raised. Liquid metals, in contrast, are good thermal conductors.

For crystalline dielectrics and some semiconductors k increases with decreasing temperature, reaching peak values at low temperatures before falling off rapidly at still lower temperatures. It should be noted that at low

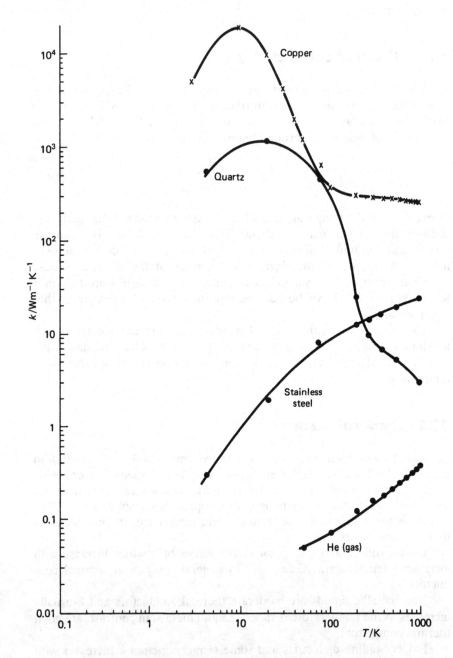

Figure 18.1 *Thermal conductivities of various substances*

temperatures some of these materials have a higher value for k than have good metallic conductors, such as copper, while at room temperature many non-metals have a higher thermal conductivity than have many metals.

These results are a consequence of the differing mechanisms of heat conduction in the various materials, and these mechanisms will now be considered briefly.

18.3 Mechanisms of heat conduction

In metals and alloys there is a strong correlation between thermal conductivity k and electrical conductivity γ, contained in the Wiedemann–Franz–Lorentz law

$$\frac{k}{\gamma T} = \text{constant} \tag{18.2}$$

where T is the temperature and the constant has the same value for all metallic conductors.

Experimental results (Table 18.1) show that, around room temperature, Equation (18.2) is a good approximation, although there is a marked falling-off in $k/\gamma T$ as the temperature approaches absolute zero. This correlation indicates that in metals and alloys heat conduction is almost entirely by the free electrons.

Table 18.1 Values of $(k/\gamma T)/10^{-8}$ V^2 K^{-2} for some common metals, to illustrate the Wiedemann–Franz–Lorentz law

Metal	Temperature				
	103 K	173 K	273 K	291 K	373 K
Copper	1.85	2.17	2.30	2.32	2.32
Silver	2.04	2.29	2.33	2.33	2.37
Zinc	2.20	2.39	2.45	2.43	2.33
Lead	2.55	2.54	2.53	2.51	2.51

Values from Roberts and Miller (1951).

In contrast, there is no correlation between γ and k for electrical insulators and semiconductors. Heat transfer in these materials is mainly by high-frequency elastic waves, known as phonons, since the energy is quantised.

Conduction by phonons must occur in all solids, but in metals the effect is greatly reduced, because the phonons are scattered by the free electrons. In general, the thermal conductivity of a solid is the sum of two parts, one resulting from conduction by vibrations of the crystal structure and the other from electron movement. Usually one effect is dominant, but in a few materials the effects are comparable.

The free electrons in a metallic crystal are scattered by imperfections in the ideal atomic arrangement. These imperfections may be, for example, impurity atoms, but also include the displacement of atoms from the ideal positions as they vibrate. The amplitude of vibration increases with temperature, so that the probability of scattering increases and the mean free path of the electrons decreases. However, the energy carried by the electrons increases with temperature, so that k, which depends on the product of mean free path and energy, is independent of temperature. Electrical conductivity depends only on the mean free path of the electrons and, therefore, decreases with temperature. Combining these two conclusions gives the Wiedemann–Franz–Lorentz law.

In non-metals k depends on the number of phonons and on their mean free path. As the temperature rises, the number of phonons increases but their mean free path decreases faster than the number of phonons rises, and k falls. As the temperature is reduced, the mean free path increases and k rises until, when the mean free path is comparable to the dimensions of the sample, the phonons are reflected back into the sample and k decreases again.

18.4 Measurement of thermal conductivity

The simplest methods for determining the thermal conductivity of a medium are the direct static (or steady-state) methods which apply the defining Equation (18.1) to a region of the sample in which the temperature distribution is independent of time.

It is not possible to measure directly the temperature gradient at any point in a material and, therefore, Equation (18.1) must be used in integral form — that is,

$$\frac{dq}{dt} \int \frac{dn}{A} = - \int_{T_1}^{T_2} k\, dT \; ; \; T_1 > T_2 \tag{18.3}$$

where dq/dt is the steady energy flux. Equation (18.3) may be written

$$\frac{dq}{dt} F = \overline{k}(T_1 - T_2) \tag{18.4}$$

where $\overline{k}$ is the average value of k in the temperature range T_2 to T_1, defined by

$$\overline{k} = \frac{1}{T_1 - T_2} \int_{T_2}^{T_1} k\,\mathrm{d}T$$

F is a 'form factor' which may be calculated explicitly in a few arrangements of simple geometry, such as those discussed below.

The simplest arrangement is that in which the flow of heat is unidirectional and perpendicular to the faces of a slab of material, as in Figure 18.2. Let the slab have an area of cross-section A and a thickness x.

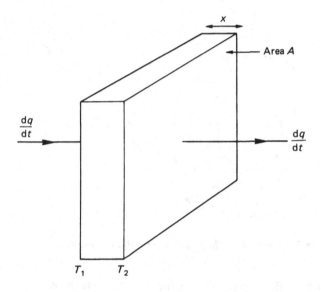

Figure 18.2 *Linear heat flow through a slab of material*

When conditions are steady, the rates of heat flow across the two faces of the slab are equal, provided that there are no heat losses from the side surfaces. Then $-\partial T/\partial n$ is constant and, if x is measured in the direction of decreasing T, is equal to $(T_1 - T_2)/x$. Equation (18.3) then becomes

$$\frac{\mathrm{d}q}{\mathrm{d}t}\frac{x}{A} = \overline{k}(T_1 - T_2) \qquad (18.5)$$

A radial heat flow may be obtained when the sample fills the space between two coaxial cylinders of radii r_1 and r_2, respectively ($r_2 > r_1$), both of length L. Let the inner surface of the sample be maintained at a constant temperature T_1 and the outer surface at a constant temperature T_2, with $T_1 > T_2$, as in Figure 18.3. When conditions are steady, the heat flow is radial (from symmetry) and the rate of heat flow over any coaxial cylindrical surface in the material is constant. Under these conditions let T be the temperature of the material at a distance r from the cylindrical axis.

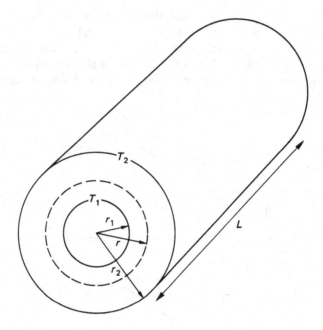

Figure 18.3 *Radial heat flow through a hollow cylinder of material*

The rate of heat flow across this surface is dq/dt, given by

$$\frac{dq}{dt} = - k \cdot 2\pi r L \frac{dT}{dr} \qquad (18.6)$$

since the area of the cylindrical surface is $2\pi r L$ and the radial temperature gradient at a distance r from the axis is dT/dr. Rearranging Equation (18.6) gives

$$k dT = - \frac{dq}{dt} \frac{dr}{2\pi r L}$$

so that

$$\int_{T_1}^{T_2} k dT = - \frac{dq}{dt} \int_{r_1}^{r_2} \frac{dr}{2\pi r L}$$

and

$$\bar{k}(T_2 - T_1) = -\frac{dq}{dt}\frac{1}{2\pi L}\ln\left(\frac{r_2}{r_1}\right)$$

or

$$T_1 - T_2 = \frac{dq/dt}{2\pi \bar{k}L}\ln\left(\frac{r_2}{r_1}\right) \tag{18.7}$$

To measure k for a solid good conductor, the arrangement shown in Figure 18.4 may be used. This is essentially that used by Lees in 1908. A specimen Sp is taken in the form of a bar, with one end maintained at a constant temperature by being placed in contact with a current-carrying resistor R, and the other at a lower temperature through contact with a heat sink H. For measurements at around room temperature H is often a

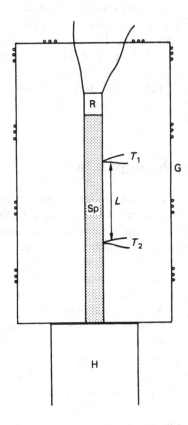

Figure 18.4 *The measurement of thermal conductivity for a solid good conductor*

water cooled coil. The surface of the specimen is either lagged or fitted with a guard tube G. This guard tube has a number of resistors attached to it, so that it can reproduce closely the temperature distribution in the specimen, and so make the heat losses very small. Thermocouples are used to measure the temperatures T_1 and T_2, respectively, at points a distance L apart along the length of the bar, and $\bar{k}$ in the temperature range T_1 to T_2 is obtained from

$$\frac{dq}{dt} = \bar{k}A \frac{(T_1 - T_2)}{L} \tag{18.8}$$

where A is the area of cross-section of the rod and dq/dt is the rate at which heat reaches the heat sink H. dq/dt is equal to $\bar{c}(T_3 - T_4)(dm/dt)$, where T_3 and T_4 are, respectively, the inlet and outlet temperatures of the water flowing through H, dm/dt is the mass flow-rate and $\bar{c}$ is the mean specific heat capacity in this temperature range.

For solid poor conductors the sample is made in the form of a thin slab (Lees, 1898). This arrangement gives a reasonable temperature gradient without the need for a very large rate of heat flow.

When k is determined for liquids and gases, heat transport by convection must be avoided. To ensure this the sample is arranged to be a thin horizontal layer and energy is supplied to the top of the layer only. The least dense fluid is then at the top of the sample and convection is prevented. When the fluid is transparent, there may be appreciable energy transfer by radiation.

18.5 Heat flow through a bar

The mathematical theory of heat conduction generalises Equation (18.1) to three dimensions and then endeavours to determine the rate of heat flow in a particular situation by solving the equation and adjusting the solution to fit the boundary conditions. This general approach will not be considered here, but some indication of the information that can be obtained in this way is gained by examining the flow of heat through a narrow bar.

Consider a long bar of uniform composition and constant area of cross-section (Figure 18.5). If the bar is initially at the temperature of the surroundings T_s and, from time zero, one end is maintained at a steady temperature T_0, there will be a period when the temperature distribution in the bar changes with time until, eventually, a steady state is achieved.

At some time t let the temperature at a plane section P_1 of the bar (see Figure 18.5), distant x from the end maintained at a temperature T_0, be T. It is assumed that the cross-section of the bar is sufficiently small for the temperature distribution in the bar to be treated as a function of x only. Then, at the same instant, the temperature at the plane section P_2 is

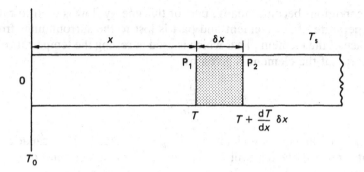

Figure 18.5 *Linear flow of heat through a long bar*

$$\left(T + \frac{dT}{dx}\delta x\right)$$

where δx is the separation of P_1 and P_2. The rate of heat flow across the plane P_1 at this instant is

$$\left(\frac{dq}{dt}\right)_{P_1} = -k_1 A\frac{dT}{dx} \tag{18.9}$$

where k_1 is the thermal conductivity of the material at a temperature T and A is the area of cross-section of the bar. At the same instant the rate of heat flow across the plane P_2 is

$$\left(\frac{dq}{dt}\right)_{P_2} = -k_2 A\frac{d}{dx}\left(T + \frac{dT}{dx}\delta x\right)$$

$$= -k_2 A\frac{dT}{dx} - k_2 A\frac{d^2T}{dx^2}\delta x \tag{18.10}$$

where k_2 is the thermal conductivity of the material at a temperature $T + [(dT/dx)\delta x]$.

The net rate of a flow of heat into the element of material between P_1 and P_2 is dq/dt, given by

$$\frac{dq}{dt} = \left(\frac{dq}{dt}\right)_{P_1} - \left(\frac{dq}{dt}\right)_{P_2}$$

If the thermal conductivity is assumed constant, so that $k_1 = k_2 = k$, say,

$$\frac{dq}{dt} = kA\frac{d^2T}{dx^2}\delta x \tag{18.11}$$

Before conditions become steady, part of this energy flow is used to raise the temperature of the element and part is lost to the surroundings from the surface of the element. That which is used in raising the temperature of the material of the element is given by

$$mc_p\frac{dT}{dt}$$

where m is the mass of the element and c_p is the specific heat capacity at constant pressure, which will be assumed to be a constant over the temperature range involved. If ρ is the density of the material

$$m = \rho A\delta x$$

It will be assumed that the rate of heat loss from unit area of the surface of the bar may be written $h(T - T_s)$, where h is a constant whose value depends on the nature of the surface. h is a heat transfer coefficient (see Section 18.9), although in the present context it is often called the *surface emissivity*. Then, if s is the perimeter of a cross-section of the bar, the rate of heat loss from the surface of the element at a temperature T is

$$hs\delta x(T - T_s)$$

The law of conservation of energy may then be written

$$kA\frac{d^2T}{dx^2}\delta x = mc_p\frac{dT}{dt} + hs\delta x(T - T_s)$$

$$= \rho A\delta x c_p\frac{dT}{dt} + hs\delta x(T - T_s)$$

and this equation may be put in the form

$$\frac{dT}{dt} = a\frac{d^2T}{dx^2} - \mu(T - T_s) \tag{18.12}$$

where

$$a = \frac{k}{\rho c_p} \text{ and } \mu = \frac{hs}{A\rho c_p}$$

a is known as the *thermal diffusivity* of the material and Equation (18.12) is known as Fourier's equation.

When conditions are steady, dT/dt is zero and Equation (18.12) becomes

$$\frac{d^2T}{dx^2} = m^2(T - T_s) \tag{18.13}$$

where

$$m^2 = \frac{\mu}{a} = \frac{hs}{Ak}$$

If the bar is thoroughly lagged, there are no heat losses and h is zero. Under these conditions Equation (18.13) becomes

$$\frac{d^2T}{dx^2} = 0 \qquad (18.14)$$

and integrating twice gives

$$T = Ax + B \qquad (18.15)$$

where A and B are constants. Therefore, when conditions are steady, the relationship between T and x in a lagged bar is linear. Further, the condition that the area of cross-section be small may then be relaxed.

For an unlagged bar of small area of cross-section the steady-state temperature distribution is obtained by solving Equation (18.13). Try a solution of the form

$$T - T_s = e^{nx}$$

where n is a constant. Then

$$\frac{dT}{dx} = ne^{nx} = n(T - T_s)$$

and

$$\frac{d^2T}{dx^2} = n^2(T - T_s)$$

Substituting for d^2T/dx^2 in Equation (18.13) gives

$$n^2(T - T_s) = m^2(T - T_s)$$

or

$$n = \pm m \qquad (18.16)$$

Therefore the complete solution is

$$T - T_s = A'e^{mx} + B'e^{-mx} \qquad (18.17)$$

where A' and B' are constants.

As an example of the application of Equation (18.17), consider the Ingen–Hausz experiment, which gives a comparison of the thermal conductivities of materials in the form of long rods.

Bars of different materials are prepared having the same dimensions and being similarly coated to give the same surface emissivity. The steady-state distribution of temperature with distance is measured for each rod when the ends at $x = 0$ are maintained at a constant temperature T_0. Let T_L be the temperature at a distance L from this end. If the bars are sufficiently long, their cold ends will be at the temperature T_s of the surroundings. This gives an effective boundary condition $T = T_s$ when $x = \infty$, which together with the condition $T = T_0$ when $x = 0$, enables A' and B' to be determined.

For the Ingen–Hausz arrangement the values obtained are

$$A' = 0; \ B' = T - T_s$$

Since $T = T_L$ when $x = L$, Equation (18.17) may be written

$$T_L - T_s = (T_0 - T_s)e^{-mL}$$

or

$$\frac{T_L - T_s}{T_0 - T_s} = e^{-mL}$$

so that

$$\ln\left(\frac{T_L - T_s}{T_0 - T_s}\right) = -mL \tag{18.18}$$

If T_L is taken to be the same for each bar, the left-hand side of Equation (18.18) is a constant so that, for bars numbered 1, 2, 3, ...,

$$m_1L_1 = m_2L_2 = m_3L_3 = \ldots \tag{18.19}$$

where L_1 is the distance along bar 1 that gives a temperature drop from T_0 to T_L. Now $m^2 = hs/Ak$ and h, s and A have been made the same for all the bars. Therefore,

$$\frac{L_1^2}{k_1} = \frac{L_2^2}{k_2} = \frac{L_3^2}{k_3} = \ldots \tag{18.20}$$

18.6 Heat conduction and entropy

The conduction of heat under a finite temperature gradient is an irreversible process and must, therefore, be accompanied by an increase in the entropy, provided that it occurs within an adiabatic boundary. A detailed examination will show where the entropy production is located.

Consider a heat reservoir at a temperature T_1 connected to another at a temperature T_2 by a lagged bar, and assume $T_1 > T_2$. When conditions are steady, the transfer of a quantity of heat Q from the higher-temperature reservoir produces a change in its entropy of $-Q/T_1$. Similarly, when this quantity of heat is absorbed by the lower-temperature reservoir, its entropy changes by Q/T_2. Since the lagged conductor is in a steady state, it suffers no change in entropy during the energy transfer. Therefore, the total change in entropy ΔS of the two reservoirs and the lagged conductor is

$$\Delta S = \frac{Q}{T_2} - \frac{Q}{T_1} = \frac{Q(T_1 - T_2)}{T_1 T_2} \tag{18.21}$$

The situation may also be analysed by considering the lagged conductor as the system. When conditions are steady, the rate of heat flow along the bar, dq/dt, is constant. The rate at which entropy flows into the bar from the reservoir at temperature T_1 is $(dq/dt \div T_1)$ and the rate at which it flows out to the reservoir at a temperature T_2 is $(dq/dt \div T_2)$. Since the bar is in a steady state, its entropy cannot change with time. Therefore, to maintain the steady state, the transfer of energy through the conductor must generate entropy at a rate given by

$$\frac{dS}{dt} = \left(\frac{dq}{dt}\right)\left(\frac{1}{T_2} - \frac{1}{T_1}\right)$$
$$= \left(\frac{dq}{dt}\right)\left(\frac{T_1 - T_2}{T_1 T_2}\right) \tag{18.22}$$

or, in the limit of vanishingly small temperature differences,

$$\frac{dS}{dt} = \left(\frac{dq}{dt}\right)\frac{dT}{T^2} \tag{18.23}$$

Calling $(dq/dt) \div T$ the *entropy current* I_S, Equation (18.23) may be written

$$\frac{dS}{dt} = I_S \frac{dT}{T} \tag{18.24}$$

Equation (18.24) shows how the entropy current builds up as heat flows along the bar — that is, it shows where the production of entropy takes place.

18.7 Heat pipes

A device in which the energy transport is dominated by evaporation and condensation is the heat pipe.

A heat pipe transfers energy from one location to another by boiling a liquid at the source of the heat, allowing the vapour to flow to the end of the pipe and condensing it there, so that the enthalpy of vaporisation is released. The liquid is then returned to the hotter end of the heat pipe by capillary action in a wick that forms the inner surface of the pipe (see Figure 18.6). Once it has returned to the hotter end of the pipe, the liquid is again vaporised, to repeat the cycle.

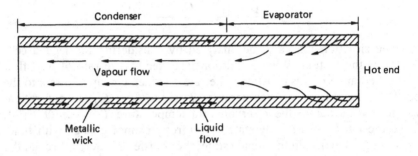

Figure 18.6 *The structure of a heat pipe*

Heat pipes are very efficient heat transfer devices. The temperatures at which a liquid boils and its vapour condenses are equal when the pressures are the same. Therefore, since the pressure in a heat pipe is almost uniform, there is only a small temperature difference between its ends. Consequently, the effective thermal conductivity can be much higher than that of the best metallic conductors. Also, as the boiling point of a liquid depends on the pressure, a heat pipe can be made to operate at a chosen temperature by using enough fluid to keep the pressure at the required value for that temperature.

18.8 Convection

Convection is the motion of the hot body carrying its energy with it and can only occur in a fluid. A current of fluid that absorbs heat in one location and then moves to another location where it mixes with cooler fluid and rejects heat is called a convection current.

When the motion of the fluid is caused by the difference in density which accompanies a difference in temperature, the heat transfer process is called natural convection. If, however, the motion of the fluid is caused primarily by the action of a pump or fan, the process is forced convection.

At low speeds the flow of a fluid is essentially in layers, without mixing on the macroscopic scale (*streamline flow*), but for speeds above a critical value there is complete mixing (*turbulent flow*). However, even when the bulk of the fluid is moving in turbulent flow, there is still a thin layer in contact with the solid containing surface in which streamline flow takes place. This region, where the transfer of energy occurs by random molecular motion only, provides most of the resistance to heat flow. The *critical speed* for the change from streamline flow to turbulent flow increases as the kinematic viscosity of the fluid increases, but decreases as the width of the channel increases, and for broad channels will be very small.

For a fluid in contact with a solid surface at a different temperature, convection currents ensure that the bulk of the fluid is at a uniform temperature. The temperature change ΔT between the solid and the bulk fluid occurs in the narrow layer of fluid close to the surface where the flow is streamline. In the case of a fluid in contact with a plane surface, the variation of temperature in the fluid, normal to the surface, is shown schematically in Figure 18.7.

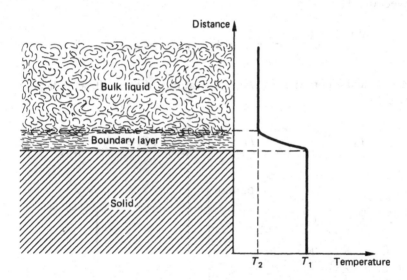

Figure 18.7 *The temperature distribution across the interface between a liquid and a solid at a higher temperature*

Heat is transferred between the solid surface and the bulk of the fluid by a combination of conduction, convection and radiation: largely by conduction in the narrow layer of fluid adjacent to the surface and largely by convection in the bulk of the fluid. The net effect of these mechanisms may be included in a heat transfer coefficient h, defined by

$$\frac{\mathrm{d}q}{\mathrm{d}t} = hA\Delta T \tag{18.25}$$

where dq/dt is the rate of heat transfer through an area A of the solid surface when the temperature difference between the solid surface and the bulk of the fluid is ΔT. The determination of h is the major aim of convection studies. As has been pointed out earlier (Appendix to Chapter 11), for ordinary laboratory apparatus, h is a constant for strongly forced convection (Newton's law of cooling), while for natural convection h is proportional to $(\Delta T)^n$, where n lies usually between $\frac{2}{3}$ and $\frac{4}{5}$.

The concept of a heat transfer coefficient for a single interface may be readily extended to more complex situations, such as the arrangement shown in Figure 18.8. Here heat is transferred from fluid 1, with a bulk temperature T_1, through a solid wall of thickness x to fluid 2, having a bulk temperature T_2 $(T_1 > T_2)$. Let the wall material have a thermal conductivity k and let the heat transfer coefficients be h_1 and h_2, respectively. When conditions are steady, the rate at which heat is conducted through an area A of the wall, dq/dt, is equal to the rate at which it is supplied by convection. Therefore,

$$\frac{dq}{dt} = kA\frac{(T_3 - T_4)}{x} = h_1A(T_1 - T_3) = h_2A(T_4 - T_2)$$

where T_3 and T_4 are the surface temperatures of the wall, as indicated in Figure 18.8. Then

$$T_3 - T_4 = \frac{x\,dq/dt}{kA};\ T_1 - T_3 = \frac{dq/dt}{h_1A}\ ;\ T_4 - T_2 = \frac{dq/dt}{h_2A}$$

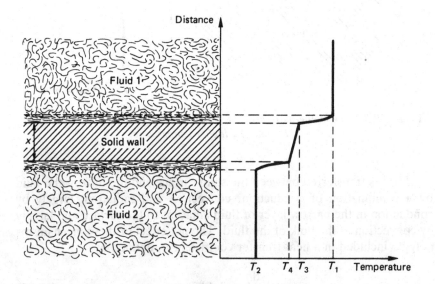

Figure 18.8 *The temperature distribution across a solid wall separating two fluids at different temperatures*

and adding these three equations gives

$$T_1 - T_2 = \frac{dq/dt}{A}\left(\frac{x}{k} + \frac{1}{h_1} + \frac{1}{h_2}\right)$$ (18.26)

Equation (18.26) may be written

$$\frac{dq}{dt} = A\overline{U}(T_1 - T_2)$$

where

$$\frac{1}{\overline{U}} = \frac{x}{k} + \frac{1}{h_1} + \frac{1}{h_2}$$

By comparison with Equation (18.25), the quantity $\overline{U}$ is called the *total heat transfer coefficient* for the particular arrangement.

18.9 Heat transfer coefficients

The calculation of theoretical values for h is extremely difficult. Even in a simple situation, such as natural convection above a hot horizontal disc in the open air, the convection current pattern is complex. Air rising vertically from the disc is replaced by air flowing in from the sides, so that, if the disc is large, the air reaching the centre has already had its temperature raised by the outer parts of the disc. Also, since air rises from the whole plate, more air enters at the edges than is needed to carry away the energy from the outer parts and, in fact, the airflow breaks up into cells in each of which air descends over the outer edges and rises from the middle. At the same time the temperature of the disc becomes non-uniform. In more complex situations the convection current pattern becomes even more complicated.

In general, the values of h must be determined experimentally and, at first sight, it would appear that this entails studying the heat loss from the chosen surface as a function of fluid velocity, temperature excess and surface size for a large number of fluids. However, much of this experimental work can be avoided by the application of dimensional analysis.

As an example of such application, consider the rate of heat loss from a hot horizontal plate with a gas streaming over it. The heat loss per unit area per unit time, H, may be considered to depend on the following quantities: fluid speed v, linear scale or size L, temperature difference between surface and fluid ΔT, fluid viscosity η, fluid density ρ, fluid thermal conductivity k, fluid specific heat capacity at constant pressure c_p, and, to account for buoyancy effects, the product of fluid cubic expansivity β and the acceleration of free fall g. If it is assumed that there is a relationship of the form

$$H \propto v^a L^b \Delta T^c \eta^d k^e \rho^f c_p^i (\beta g)^j$$

where a, b, ..., j are constants, dimensional analysis shows that

$$\frac{HL}{k\Delta T} = \frac{hL}{k} \propto (Re)^x (Pr)^y (Gr)^z \tag{18.27}$$

where x, y, z are constants appropriate to a particular size of the plate and (Re), (Pr) and (Gr) are dimensionless groups defined as follows:

$(Re) = v\rho L/\eta$, the *Reynolds number*.
$(Pr) = c_p \eta/k$, the *Prandtl number*.
$(Gr) = \beta g \, \Delta T L^3 \rho^2/\eta^2$, the *Grashof number*.
hL/k is often called the *Nusselt number*, (Nu).

The precise form of the relationship between (Nu), (Re), (Pr) and (Gr) must be found by experiment once it has been shown that the dimensionless groups are consistent with the measurements. However, further simplifications to Equation (18.27) are possible. In natural convection the forced speed v may be dropped from the analysis. This is equivalent to putting x equal to zero to eliminate (Re), giving, under these conditions,

$$(Nu) = \text{function of } (Pr) \text{ and } (Gr)$$

Under conditions of forced convection the effects of buoyancy are, in general, negligible and Equation (18.22) may then be written

$$(Nu) = \text{function of } (Re) \text{ and } (Pr)$$

For many gases (Pr) has nearly the same value over a wide range of temperatures and may be treated as a constant. Then, for natural convection in gases,

$$(Nu) = \text{function of } (Gr)$$

while for forced convection in gases

$$(Nu) = \text{function of } (Re)$$

Experimentally, this last relationship, for example, could be tested by measuring the heat transfer from a hot horizontal plate to a gas as a function of gas speed, say. If a unique curve is obtained when (Nu) is plotted against (Re), the validity of the assumptions is confirmed and the heat transfer may be obtained from the graph under any conditions in the range of (Re) studied. The curve can be used for gases of different viscosity and density and also for discs of different size, but separate curves are needed for different geometrical arrangements.

18.10 Thermal radiation

Thermal radiation is the electromagnetic radiation a body emits by virtue of its temperature. The medium through which the radiation is passing suffers no change in its properties (in temperature, for example) unless some of the radiation is absorbed. These properties of thermal radiation are embodied in Maxwell's definition of the process: in radiation the hot body loses energy and the cold body gains energy by some process occurring in the intervening medium, which does not itself thereby absorb energy.

It is usually assumed that gases are transparent to thermal radiation, and certainly gases with symmetric molecules (e.g. H_2, O_2, N_2) do not radiate appreciably, even at high temperatures, nor do they absorb radiation passing through them. Gases with heteropolar molecules (e.g. H_2O, CO_2) may absorb strongly within certain limited wavelength bands characteristic of the particular gas.

When thermal radiation is dispersed, a continuous spectrum is obtained. The distribution of energy among the various wavelengths depends on the temperature of the radiating body and the nature of its surface. At temperatures below about 800 K most of the energy is associated with infra-red waves, whatever the nature of the surface, whereas at higher temperatures the region of strongest emission moves to shorter wavelengths. In general, the higher the temperature of the radiating system the greater the total energy radiated.

When a system is placed in a vacuum, the only way in which it can receive energy is by the absorption of radiation from its surroundings. The internal energy of a system and, therefore, its temperature will remain constant when the rate at which it absorbs energy is equal to the rate at which its emits energy. This is the essence of *Prévost's theory of exchanges*, which states that, when a uniform temperature is reached in a system of several bodies, although conduction and convection must cease, radiation does not. Rather, each body is radiating to the others and receiving radiation from them, but the net effect is zero. It follows that the rate of emission of thermal radiation from a surface is independent of the surroundings.

The total radiant energy emitted by a body per unit area per unit time is known as the *radiant emittance R* of the surface. R depends on the nature of the surface and the temperature. For a given surface it will be written $R(T)$. *Radiant flux* $\varphi(T)$ is the total energy radiated per unit time by a given surface.

Since the radiant energy emitted by a surface is not distributed uniformly with wavelength, it is convenient to define the *spectral emissive power* $e_\lambda(T)$ at a given temperature and wavelength λ, so that the radiant energy emitted per unit area per unit time between wavelengths λ and $\lambda + d\lambda$ is $e_\lambda d\lambda$. Then, for a given surface,

$$\int_0^\infty e_\lambda(T)\mathrm{d}\lambda = R(T) \tag{18.28}$$

In general, a surface absorbs only a fraction of the thermal radiation that falls on it. The *absorptance* or *absorption factor* $\alpha(T)$ is the fraction of the total radiant energy of isotropic radiation falling upon a surface which is absorbed at that temperature and, for a given surface, is a function of temperature. Radiation is termed isotropic when it is incident upon a surface equally from all directions. If φ_0 is the radiant flux falling on a surface that is at a temperature T, and φ_a is the flux that is absorbed, $\alpha(T) = \varphi_a/\varphi_0$.

Similar definitions follow for *reflectance* or *reflection factor* $\rho(T)$ $(= \varphi_r/\varphi_0)$ and *transmittance* or *transmission factor* $\tau_{tr}(T)(= \varphi_{tr}/\varphi_0)$, where φ_r is the flux reflected and φ_{tr} that transmitted. From the conservation of energy,

$$\alpha + \rho + \tau_{tr} = 1$$

The fraction of the radiation absorbed is also, in general, a function of wavelength. *Spectral absorptance* $\alpha_\lambda(T)$, for a given wavelength and temperature, is that fraction of the isotropic radiation lying between wavelengths λ and $\lambda + \mathrm{d}\lambda$ that is absorbed at that temperature.

18.11 Black-body radiation

The concept of the black body is essentially an extrapolation of the observed behaviour of real bodies: a *black body* is defined as one that is capable of absorbing all the thermal radiation falling on it at all temperatures. Therefore, the absorptance of a black body α_B is unity.

A perfect black body cannot be realised in practice, but, as will be shown later in this section, the radiation emitted by a black body at a given temperature is identical with that existing in an evacuated, closed cavity, with opaque walls at the same temperature as that of the black body. The cavity walls emit thermal radiation which is then partially absorbed and partially reflected a large number of times, with the result that the cavity is filled with isotropic radiation. A useful measure of the energy of this cavity radiation is the *irradiance E* within the cavity. This is defined as the energy falling in unit time on unit area of any surface placed within the cavity. When the walls of the cavity are opaque, E is a function of the temperature only, and does not depend on the nature of the walls. The same is true of the spectral distribution of the radiation. Consider two cavities A and B, having opaque walls maintained at the same temperature. Let A and B be connected by a narrow tube along which radiation may pass but which does not disturb the radiation in either cavity when communication is effected. If the irradiance in A is greater than that in B, when the communicating tube is opened the rate of energy transfer from A to B is greater than that from B to A. Consequently, there is a net energy transfer between two systems at the same temperature, a result which violates the Clausius statement of the second law of thermodynamics. Therefore, whatever the nature of the opaque walls, the irradiance in A must equal that in B and be a function of the temperature only — that is, $E = E(T)$.

If the connecting tube is fitted with a filter that passes only radiation in the wavelength range λ to $\lambda + d\lambda$, a repetition of the above argument shows that the spectral distribution of the energy within the cavity must be a function of temperature only. Further, if the connection is made at different positions on the cavities, it may be shown in a similar manner that the radiation within each cavity is isotropic.

The relation between the radiant emittance of a black body and the irradiance within a cavity may be found by placing a black body at a temperature T into an evacuated cavity, the walls of which are maintained at the same temperature. In this arrangement the temperature of the black body is unchanged, so that the rate at which it absorbs energy from the cavity radiation must be equal to the rate at which it emits radiation. The radiant energy emitted per unit area per unit time by the black body is R_B and that absorbed per unit area per unit time is $\alpha_B E(T)$, which is equal to $E(T)$, since α_B is unity. Therefore,

$$R_B = E(T) \tag{18.29}$$

i.e. the irradiance within an evacuated cavity whose opaque walls are maintained at a temperature T is equal to the radiant emittance of a black body at the same temperature. For this reason the radiation within a cavity is often referred to as *black-body radiation*. Since E is a function of T only, it follows that R_B is a function of T only.

Black-body radiation may be studied by allowing radiation to escape from a very small hole in the cavity wall. The hole must be so small that its presence does not disturb the radiation in the cavity. This hole acts as a black body, since it emits thermal radiation of the same quantity and quality (spectral distribution) as a black body at the same temperature as that of the cavity wall, and all radiation falling on the hole is completely absorbed within the cavity by successive reflection at the walls.

18.12 The radiation laws

The radiant emittance of a non-black body depends on the nature of the surface as well as on the temperature. Let a non-black body at a temperature T be introduced into a cavity, the walls of which are also at a temperature T. If the irradiance within the cavity is E and the absorptance of the body is $\alpha(T)$, the radiant energy absorbed by the body per unit area per unit time is $\alpha(T)E(T)$. The energy that the body emits per unit area per unit time is the radiant emittance $R(T)$. Since the non-black body is in equilibrium in the radiation field within the cavity

$$R(T) = \alpha(T)E(T)$$

But, from Equation (18.29),

$$E(T) = R_B(T)$$

so that

$$R(T) = \alpha(T)R_B(T) \tag{18.30}$$

Equation (18.30) states that the radiant emittance of a non-black body at a given temperature is equal to a fraction of the radiant emittance of a black body at the same temperature, the fraction being the absorptance at that temperature. This is Kirchhoff's radiation law. The non-black body emits radiation that has a quality and quantity such that, when combined with the reflected radiation, it produces an isotropic distribution of radiation from the surface which is identical with $E(T)$. The ratio R/R_B — that is, the absorptance α — is also known as the *emissivity* ϵ of the surface. Kirchhoff's law shows that, at a given temperature, a good absorber of thermal radiation is also a good emitter of thermal radiation.

It is instructive to develop this law in terms of spectral absorption and emission. Let the irradiance within the cavity for the wavelength range λ to $\lambda + d\lambda$ be E_λ. Then, when the non-black body is in equilibrium,

$$\alpha_\lambda E_\lambda = e_\lambda d\lambda$$

or

$$\frac{e_\lambda}{\alpha_\lambda} = \frac{E_\lambda}{d\lambda} \tag{18.31}$$

where α_λ is the spectral absorptance. Now $E_\lambda/d\lambda$ depends only on the wavelengths of the radiation and the temperature of the cavity walls. Therefore, e_λ/α_λ is a universal function of wavelength and temperature for all surfaces.

Consider now a cavity with walls at a constant temperature T_W and let a non-black body at a different temperature T be placed in the cavity. This body is not in equilibrium in the radiation field within the cavity and will not radiate energy of the same quality and quantity as that which it absorbs. For a simple analysis of the situation let the non-black body be so small that introducing it into the cavity does not significantly affect the character of the radiation within the cavity. The rate at which radiant energy is absorbed per unit area by the body is $\alpha(T)E(T_W)$ and the rate at which it is emitted per unit area is $R(T)$. These two rates are not equal; the difference between them is the net rate of energy transfer, per unit area of the body, between the body and the cavity walls. This energy transfer arises because T is not equal to T_W and is, therefore, a heat flow. If the surface area of the non-black body is A, the net rate of heat transfer between the body and the cavity walls is dq/dt, given by

$$\frac{dq}{dt} = \alpha(T)E(T_W)A - R(T)A \qquad (18.32)$$

Now

$$E(T_W) = R_B(T_W) \text{ (Equation 18.29)}$$

and

$$R(T) = \alpha(T) R_B(T) \text{ (Equation 18.30)}$$

Therefore,

$$\frac{dq}{dt} = A\alpha(T)[R_B(T_W) - R_B(T)] \qquad (18.33)$$

When the body is a black body, $\alpha(T)$ is unity and

$$\frac{dq_B}{dt} = A[R_B(T_W) - R_B(T)] \qquad (18.34)$$

The first measurements of the energy transfer by radiation between a body and its surroundings were made by Tyndall and published in 1863. On the basis of these measurements, Stefan concluded in 1879 that the net energy radiated was proportional to the difference between the fourth powers of the thermodynamic temperatures. Then, in 1884, Boltzmann showed theoretically (see Section 18.14) that the radiant emittance of a black body at a temperature T is given by

$$R_B(T) = \sigma T^4 \qquad (18.35)$$

Equation (18.35) is the Stefan–Boltzmann law and σ is the *Stefan–Boltzmann constant*, equal to 5.671×10^{-8} W m^{-2}K^{-4}. Substituting for $R_B(T)$ in Equation (18.33) gives, for the heat transfer between a small non-black body at a temperature T and the walls of a cavity or enclosure at a temperature T_W,

$$\frac{dq}{dt} = A\alpha(T)\sigma(T_W^4 - T^4) \qquad (18.36)$$

The distribution of energy with wavelength for black-body radiation may be determined by examining the radiation emerging from a small hole in the wall of a constant temperature cavity at the same temperature. A convenient representation of this distribution is the graph of the spectral emissive power of a black body $e_{\lambda B}$ against wavelength λ, which is shown in Figure 18.9 for two different temperatures. Each curve passes through the origin, tends to zero for very large values of λ and is completely contained by all curves corresponding to higher temperatures. The area between any curve and the λ-axis varies as T^4, in agreement with the Stefan–Boltzmann law. If λ_m is the wavelength at which $e_{\lambda B}$ has a maximum value for a black body at a temperature T, then

$$\lambda_m T = \text{constant} = 2.89 \times 10^{-3} \text{ m K} \tag{18.37}$$

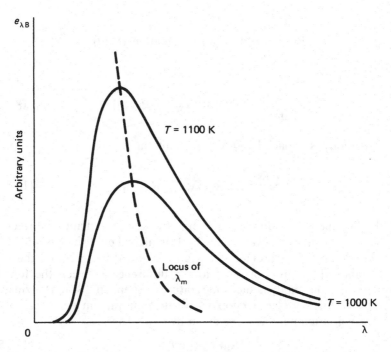

Figure 18.9 *The spectral emissive power of a black body $e_{\lambda B}$ as a function of wavelength λ for different temperatures. The dotted curve gives the locus of the maxima. Taken from page 491 (Figure 20.8) of* Heat and Thermodynamics, *by J. K. Roberts and A. R. Miller, Blackie, London and Glasgow, 1951*

Equation (18.37) is the Wien displacement law. This law may be deduced thermodynamically, as may the Wien distribution law,

$$e_{\lambda B} = C_1 \lambda^{-5} f(\lambda T) \tag{18.38}$$

where C_1 is a constant and f is a universal function of λT. However, the law giving $e_{\lambda B}$ as an explicit function of T requires the application of statistical principles together with quantum theory. This was first achieved by Planck who, in 1900, obtained the equation

$$e_{\lambda B} = \frac{8\pi hc}{\lambda^5(e^{hc/k\lambda T}-1)} \tag{18.39}$$

where h is Planck's constant, c is the speed of light in a vacuum and k here is Boltzmann's constant, not a thermal conductivity. Equation (18.39) is in complete agreement with experiment, and yields both Wien's displacement and distribution laws and the Stefan–Boltzmann law.

For real bodies $\alpha(T)$ is less than unity and α_λ varies with λ. For some surfaces α_λ also shows a directional dependence. A model surface that is useful for many materials is the *grey surface*. This has $\alpha(T)$ less than unity but α_λ independent of λ.

18.13 Determination of the Stefan–Boltzmann constant

The most precise determinations of the Stefan–Boltzmann constant σ are made using the radiation emerging from a small hole in the wall of a cavity. However, a straighforward determination of reasonable accuracy may be made by suspending a blackened metal cylinder or sphere, containing an embedded resistor, inside an evacuated enclosure or cavity. Let a current I flow through the resistor under a potential difference E. When equilibrium is achieved, the net rate of transfer of radiant energy from the black body is equal to the rate at which electrical work is done on the resistor. If the steady temperature of the wall of the enclosure is T_W and the equilibrium temperature of the black body is T, then, from Equations (18.34) and (18.35),

$$IE = A\sigma(T^4 - T_W^4) \tag{18.40}$$

where A is the surface area of the black body.

18.14 The thermodynamics of radiation

The general principle underlying the application of thermodynamics to radiation in a cavity is that the space containing the radiation may be treated as a thermodynamic system: the radiation exerts a pressure on the walls, can be expanded or compressed and can exchange energy with the surroundings.

It has already been shown (Section 18.11) that the irradiance within a cavity depends only on the temperature of the walls, so that, when a cavity is expanded quasistatically at constant temperature, new radiation is created of the same quality as that already existing.

To specify the work interaction of the radiation, it is necessary to relate the pressure exerted by the radiation to the *energy density u*. This is done most simply by treating the radiation as a photon gas and, since the photons may be regarded as non-interacting, using ideas from the simple kinetic theory of gases of Chapter 17. Since all the photons have a speed c, the pressure p exerted by the radiation may be written

$$p = \frac{1}{3}\rho c^2 \tag{18.41}$$

where ρ is the particle density (mass per unit volume) of the photons. According to Einstein's mass–energy equation, ρc^2 is the energy density in the gas, so that the pressure exerted by the photon gas in equilibrium is given by

$$p = \frac{1}{3}u \tag{18.42}$$

If the volume of the cavity containing the radiation is V, the total energy in the cavity U is given by

$$U = uV \tag{18.43}$$

and differentiating with respect to V, with T held constant, gives

$$\left(\frac{\partial U}{\partial V}\right)_T = \left(\frac{\partial u}{\partial V}\right)_T V + u \tag{18.44}$$

The photon gas is an example of a closed hydrostatic system and, for such systems, $(\partial U/\partial V)_T$ is given by the energy equation (Section 9.1) — that is, by

$$\left(\frac{\partial U}{\partial V}\right)_T = T\left(\frac{\partial p}{\partial T}\right)_V - p \tag{18.45}$$

Combining Equations (18.44) and (18.45) gives

$$T\left(\frac{\partial p}{\partial T}\right)_V - p = \left(\frac{\partial u}{\partial V}\right)_T V + u \tag{18.46}$$

Now u is a function of temperature only, so that

$$\left(\frac{\partial u}{\partial V}\right)_T = 0$$

Further,

$$p = \frac{1}{3}u$$

and, therefore,

$$\left(\frac{\partial p}{\partial T}\right)_V = \frac{1}{3}\left(\frac{\partial u}{\partial T}\right)_V$$

or, since u is a function of temperature only,

$$\left(\frac{\partial p}{\partial T}\right)_V = \frac{1}{3}\frac{du}{dT}$$

Substituting these three results in Equation (18.46) gives

$$\frac{T}{3}\frac{du}{dT} - \frac{u}{3} = u$$

or

$$4u = T\frac{du}{dT} \tag{18.47}$$

Integrating Equation (18.47) gives

$$\ln u = 4 \ln T + \text{constant}$$

or

$$u = A'T^4 \tag{18.48}$$

where A' is a constant. Equation (18.48) is essentially the Stefan–Boltzmann law.

Consider now the emission of radiation from a black body. Such a body absorbs all the radiation falling on it. When a black body at a temperature T is placed in a cavity with walls at the same temperature, to maintain equilibrium the black body must emit radiation of the same quantity and quality as that falling on it. Treating the radiation as a photon gas, it follows from Section 17.4 that the number of photons striking unit area of the black body in unit time is $nc/4$, where n is the number of photons per unit volume. Therefore, the energy absorbed by the black body per unit area per unit time is $mnc^3/4$, where m is the mass equivalent of a photon. Now the energy density u is mnc^2, so that the energy absorbed per unit area per unit time is $uc/4$. This must also be the energy emitted per unit area per unit time by the black body — that is, the radiant emittance $R_B(T)$. Therefore,

$$R_B(T) = \frac{uc}{4} \qquad (18.49)$$

But, from Equation (18.48), $u = A'T^4$. Therefore,

$$R_B(T) = \frac{A'cT^4}{4} = \sigma T^4 \qquad (18.50)$$

This is the Stefan–Boltzmann law and the Stefan–Boltzmann constant σ is $A'c/4$.

The radiation in a cavity may also be allowed to undergo an isentropic process. To isolate the radiation from its surroundings it is placed inside a cylinder having perfectly reflecting walls. This raises the problem of what is meant by the temperature of the radiation. To overcome this difficulty a minute black body is placed in the cylinder and its equilibrium temperature is taken as the temperature of the radiation. Let the radiation have the quality corresponding to that in a cavity with walls at a temperature T. It is now necessary to enquire whether or not the radiation still has the quality of black body radiation when it is expanded isentropically. Assume that, after an isentropic process, the radiation no longer has the quality of black-body radiation corresponding to the new equilibrium temperature. In particular, assume that a wavelength λ_1 is deficient and λ_2 is in excess. Now introduce into the cylinder two specks of coloured material that, respectively, emit (and absorb) selectively the wavelengths λ_1 and λ_2. The first speck emits radiation faster than it absorbs it and, therefore, becomes cooler. Similarly, the second speck becomes hotter. The temperature difference so produced can be used to drive a cyclic engine and produce work. However, this work is produced from a single reservoir, which is contrary to the second law of thermodynamics. Therefore, the process is impossible and the radiation must have been black-body radiation after the isentropic change in volume.

Treating the radiation as a closed hydrostatic system, the first law of thermodynamics is

$$dU = TdS - pdV$$

where S is the entropy of the system. When the change of volume is isentropic, dS is zero and

$$dU = -pdV$$

Now $U = uV$ and $p = u/3$, so that

$$dU = d(uV) = -\frac{1}{3}udV$$

or

$$u\mathrm{d}V + V\mathrm{d}u = -\frac{1}{3}u\mathrm{d}V$$

Therefore,

$$4\frac{\mathrm{d}V}{V} = -\frac{3\mathrm{d}u}{u}$$

or

$$V^4u^3 = \text{constant} \tag{18.51}$$

But, from Equation (18.48), $u = A'T^4$. Therefore, when black-body radiation undergoes an isentropic change in volume,

$$VT^3 = \text{constant} \tag{18.52}$$

18.15 The detection of thermal radiation

Detectors of radiant energy may be divided into two groups: (1) *thermal detectors,* which utilise the change in temperature of a system in which the radiation is absorbed, and (2) *photodetectors,* which use the quantum photoelectric effect.

Photodetector methods include the *photographic effect, photoemissive cells* and *photoconductive cells.* Such detectors count the number of effective quanta absorbed, since the action is based on the freeing of a bound electron by the absorption of a single quantum of radiation. For photocells the response time is about 10^{-6} s. The major limitation of such devices is that they do not have a uniform response at all frequencies and, in particular, each material has a cut-off frequency below which no photoelectric effect is obtained. For example, lead sulphide, deposited onto a quartz substrate, has a peak spectral response for a wavelength of $2.0 - 2.3$ μm and a spectral response range of $0.3 - 2.8$ μm.

The response of thermal detectors depends on the total energy absorbed in unit time, and the response time is typically about 10^{-3}s. Included in this group are the *bolometer* and the *thermopile.*

The bolometer, introduced by Langley in 1881, makes use of the change of resistance with temperature of a suitable material. For a high sensitivity the detecting element, which is usually a metal or a thermistor, has a small heat capacity and is blackened to increase absorption.

The thermopile, invented by Nobili and improved by Melloni, consists of a number of thermocouples connected in series. Alternate junctions are attached to a thin sheet of blackened copper which is exposed to radiation.

The other junctions are attached to a more massive sheet and shielded from radiation. In use the instrument is often fitted with a cone which concentrates the radiation on the blackened junctions and screens them from other sources of radiation.

18.16 Pyrometry

Temperatures above about 1200 K are frequently measured by instruments called *pyrometers,* which examine the radiation emitted by a body. There are three principal ways in which the temperature of a body may be determined from the thermal radiation that it emits:

1   The wavelength λ_m which gives the maximum rate of emission is found; T is then obtained from the Wien displacement law.
2   The total radiant flux from the body is found; T is obtained from the Stefan–Boltzmann law.
3   The spectral distribution of energy is found; T is obtained from Planck's law.

When the source of radiation is a black body, all three methods give the same value for the temperature of a given body, but for other bodies an *equivalent black-body temperature* is obtained which depends on the method used. For non-black-bodies the equivalent black-body temperature is always less than the true temperature, except for method 1 when it is applied to a grey surface.

Total radiation pyrometers are based on method 2 and compare the total radiation emitted by a surface in unit time with that emitted by a black body of equal area. In the instrument designed by Fery, the radiation from the body is focused onto a thermocouple by means of a concave mirror. The pyrometer is calibrated in terms of the radiation from a black body, so the emissivity must be known if the true temperature of a non-black body is to be obtained.

Disappearing filament pyrometers are based on method 3. A filament carrying an electric current has its temperature adjusted until it disappears against a background of the image of the surface whose temperature is being measured. A filter passing only a narrow range of wavelengths is used to eliminate difficulties arising from differences in colour of the surface and filament (see Figure 18.10). The instrument is calibrated using a black body, so that, when a non-black body is examined, a correction must be applied. If ϵ_λ is the spectral emissivity of the non-black body at the wavelength λ at which the pyrometer effectively acts,

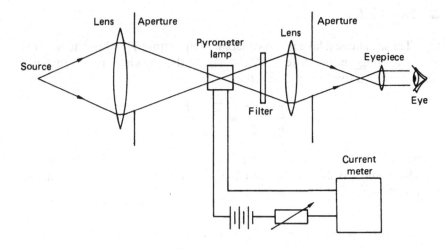

Figure 18.10 *The disappearing-filament pyrometer*

$$\epsilon_\lambda = \frac{e^{c_2/\lambda T}-1}{e^{c_2/\lambda T_B}-1} \tag{18.53}$$

where $c_2 = hc/k$ (see Equation 18.39), T_B is the temperature registered by the pyrometer and T is the true temperature of the surface of the body. The need to know ϵ_λ is the chief factor limiting the accuracy of the instrument.

If two filters are incorporated, two equivalent black-body temperatures T_1 and T_2 are obtained, corresponding, respectively, to wavelengths λ_1 and λ_2. Provided that ϵ_λ has the same value for the two wavelengths, the true temperature T is given by

$$\frac{1}{T} = \frac{1}{T_1} - \frac{\lambda_1}{\lambda_1 - \lambda_2}\left(\frac{1}{T_1} - \frac{1}{T_2}\right) \tag{18.54}$$

A development of this pyrometer is the *optical fibre thermometer*. In this a sapphire fibre has a small length at one end coated with a black material and this blackened part is placed in the region where the temperature is to be measured. When equilibrium is achieved, the black material radiates black-body radiation appropriate to its temperature along the fibre (light pipe) and the spectral distribution gives the temperature in the region of the blackened end. This technique is useful for measuring furnace temperatures.

■ EXAMPLE

Q. The air above a lake is at a constant temperature T_a, while the water is at its freezing temperature T_i ($T_a < T_i$). If the thickness of ice that has formed after a time t is y, show that

$$\frac{y}{h} + \frac{y^2}{2k} = \frac{T_i - T_a}{\rho l} t$$

where h is the heat transfer coefficient per unit area for the ice–air interface, and k is the thermal conductivity of ice, l its specific enthalpy of fusion and ρ its density.

A. Let the thickness of the ice layer increase by an amount dy in a time dt. The mass of ice formed per unit area of surface in this time is ρdy, corresponding to a release of energy, the enthalpy of fusion, of $\rho l dy$. Therefore, the rate of heat transfer from the water to the air per unit area of surface, dq/dt, is given by

$$\frac{dq}{dt} = \rho l \frac{dy}{dt}$$

The rate of heat transfer, per unit area, through the ice is given by

$$\frac{dq}{dt} = \frac{k(T_i - T_s)}{y}$$

where T_s is the temperature of the interface between the ice and the air. The rate of heat transfer per unit area from the ice to the air is given by

$$\frac{dq}{dt} = h(T_s - T_a)$$

To eliminate T_s, which is not known, write the last two equations as

$$\frac{y}{k} \frac{dq}{dt} = T_i - T_s$$

and

$$\frac{1}{h} \frac{dq}{dt} = T_s - T_a$$

Adding these equations gives

$$\left(\frac{y}{k} + \frac{1}{h}\right) \frac{dq}{dt} = T_i - T_a$$

Now substituting for dq/dt gives

$$\left(\frac{y}{k} + \frac{1}{h}\right) \rho l \frac{dy}{dt} = T_i - T_a$$

which may be integrated to give

$$\frac{y}{h} + \frac{y^2}{2k} = \frac{T_i - T_a}{\rho l} t$$

since $y = 0$ when $t = 0$.

■ **EXERCISES**

1   A furnace wall is constructed of two layers in contact, each having the same
    thickness. The temperature of the inner surface of the wall is 373 K and
    that of the outer surface is 273 K. If the thermal conductivity of the inner
    layer is nine times greater than that of the outer layer, calculate the
    temperature of the interface.

2   A thin-walled cylinder of radius R is filled with a liquid that is maintained
    at a temperature T_1. The cylinder is insulated from the surroundings by a
    layer of a poorly conducting material having a thermal conductivity k. If the
    surroundings consist of the atmosphere at a temperature T_0, show that the
    rate of loss of energy to the surroundings increases as the thickness of the
    layer is increased, until the outer radius of the layer reaches a value equal
    to k/h, where h is the heat transfer coefficient for the poorly conducting
    layer–air interface.

3   Heat flows radially through material contained between coaxial cylindrical
    surfaces of radii r_1 and r_2 ($r_2 > r_1$). The inner surface is maintained at a
    temperature T_1 and the outer surface at a temperature T_2. Determine the
    value of the radial distance r_0 in the material for which the temperature is
    $(T_1 + T_2)/2$.

4   Sir William Herschel examined the prismatic spectrum of the radiation from
    the sun, using a sensitive mercury-in-glass thermometer as the detector, and
    obtained the greatest rise in temperature beyond the red end of the visible

spectrum, in the region known as the infra-red. The sun may be treated as a black body with a surface temperature of 6000 K. Consequently, its maximum spectral emissive power is near the middle of the visible spectrum. How are these results reconciled?

5   Derive an expression for the heat capacity at constant volume of black-body radiation.

6   Show that, when a small body at a temperature T is placed in a large evacuated cavity with walls at a constant temperature T_W, the rate of heat transfer by radiation is proportional to $(T - T_W)$ when the difference between T and T_W is small.

7   Show that, when $c_2/\lambda T$ in Equation (18.53) is greater than about 5, the equation may be replaced, with negligible error, by the equation

$$\frac{1}{T} = \frac{1}{T_B} + \frac{\lambda}{c_2} \ln \epsilon_\lambda$$

Appendix: Theorems on partial differentiation

Classical thermodynamics is concerned with the equilibrium states of systems, and such states are characterised by the values of a small number of macroscopic quantities, the thermodynamic coordinates. One of the principal aims of thermodynamics is to derive relationships between the values of these coordinates for different equilibrium states. For simple systems of the kinds discussed in this book, equilibrium states are characterised by the values of two independent coordinates, and each equilibrium state is represented by a point in a three-dimensional space, using the values of the two independent coordinates and one other. The totality of equilibrium states for such a system is represented by a continuous and single-valued surface in the appropriate three-dimensional space. For example, the equilibrium states achieved by a gaseous system are represented by a continuous surface in a space, using pressure p, volume V and temperature T as coordinates. Any change that is composed of a succession of equilibrium states is represented by a line on this surface and, for an infinitesimal part of any such change, changes in the values of the coordinates are obtained by the use of the partial differential calculus. Two widely used theorems in this calculus will now be considered briefly.

Consider first a continuous and single-valued function

$$y = f(x) \qquad\qquad (A.1)$$

The differential coefficient at $x = a$ of the function $y = f(x)$ is written as $f'(a)$ and defined by the equation

$$f'(a) = \lim_{x \to a} \left[\frac{f(x) - f(a)}{x-a} \right] \qquad\qquad (A.2)$$

The function $f'(x)$, whose value at $x=a$ is $f'(a)$, is called the derivative of the function $f(x)$. Putting $x-a = h$, the differential coefficient at $x = a$ becomes

$$f'(a) = \lim_{h \to 0} \left[\frac{f(a + h) - f(a)}{h} \right] \tag{A.3}$$

The tangent to the curve at the point $x = a$, $y = b$ is the limiting position of the chords through that point. Therefore, the tangent at the point (a,b) on the curve has a gradient $f'(a)$.

If the points (a,b) and $(a + h, b + k)$ are both on the curve $y = f(x)$, then

$$\frac{k}{h} = \frac{f(a + h) - f(a)}{h} \tag{A.4}$$

When the two points are close together, h may be treated as a small increment in x and written δx; k may be treated as a small increment in y and written δy. Then

$$f'(x) = \lim_{\delta x \to 0} \left(\frac{\delta y}{\delta x} \right)$$

which is, in the usual notation,

$$f'(x) = \lim_{\delta x \to 0} \left(\frac{\delta y}{\delta x} \right) = \frac{dy}{dx} \tag{A.5}$$

For a continuous function, $f(a + h) - f(a)$ may be written

$$f(a + h) - f(a) = hf'(a) + h\epsilon(h) \tag{A.6}$$

where $\epsilon(h)$ has the property that $\epsilon(h) \to 0$ as $h \to 0$.

Now $f(a + h) - f(a)$ is an increment in f which, when h is small, may be written δf. This increment in f approaches the value $hf'(a)$ in the limit as h tends to zero. Then, if $hf'(a)$ is written df, Equation (A.6) becomes

$$\delta f = df + h\epsilon(h) \tag{A.7}$$

df is known as the differential of $f(x)$ at $x = a$ and is a multiple of $f'(a)$, the differential coefficient at $x = a$.

In particular, if $f(x)$ is the function x, $f'(a)$ is unity and h is dx, so that

$$df = f'(a) \, dx \tag{A.8}$$

Equation (A.8) gives, approximately, the change in $f(x)$ arising from a small change in x (see Figure A.1), the error being $h\epsilon(h)$.

If f is a real single-valued function of two independent variables x and y, the partial derivative of $f(x,y)$ with respect to x, written $f_x(x,y)$ or $(\partial f/\partial x)_y$, is defined as

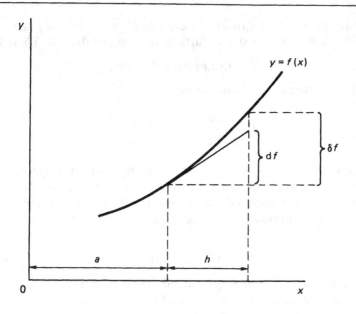

Figure A.1 *Illustrating the differential df of the function* y = f(x)

$$f_x(x,y) = \left(\frac{\partial f}{\partial x}\right)_y = \lim_{\delta x \to 0}\left[\frac{f(x + \delta x,y) - f(x,y)}{\delta x}\right] \qquad \text{(A.9)}$$

Similarly, the partial derivative with respect to y is

$$f_y(x,y) = \left(\frac{\partial f}{\partial y}\right)_x = \lim_{\delta y \to 0}\left[\frac{f(x,y + \delta y) - f(x,y)}{\delta y}\right] \qquad \text{(A.10)}$$

Extending the analysis leading to Equation (A.8) to a function of two variables $f(x,y)$ gives

$$f(a + h, b + k) - f(a,b) = hf_x(a,b) + kf_y(a,b) + \rho\epsilon \qquad \text{(A.11)}$$

where ϵ has the property that $\epsilon \to 0$ as $\rho \to 0$. Equation (A.11) may be written

$$\delta f = df + \rho\epsilon \qquad \text{(A.12)}$$

where

$$\delta f = f(a + h, b + k) - f(a,b)$$

and

$$df = hf_x(a,b) + kf_y(a,b)$$

Now, if $f(x,y) = x$, then $f_x(a,b) = 1$ and $f_y(a,b) = 0$, while, if $f(x,y) = y$, $f_y(a,b) = 1$ and $f_x(a,b) = 0$ and, further, $dx = h$ and $dy = k$. Therefore,

$$df = f_x(a,b) \, dx + f_y(a,b) dy$$

or, in the notation favoured in this book,

$$df = \left(\frac{\partial f}{\partial x}\right)_y dx + \left(\frac{\partial f}{\partial y}\right)_x dy \qquad (A.13)$$

Equation (A.13) may be used to estimate the change in f arising from small changes in x and y, the accuracy increasing as the changes tend to zero.

For a particular thermodynamic system with two independent coordinates, let the coordinates chosen be x, y and z, related by the equation of state

$$F(x,y,z) = 0 \qquad (A.14)$$

If F is a continuous and single-valued function, x may be expressed explicitly as a function of y and z — that is

$$x = x(y,z)$$

The differential of x may then be written

$$dx = \left(\frac{\partial x}{\partial y}\right)_z dy + \left(\frac{\partial x}{\partial z}\right)_y dz \qquad (A.15)$$

A similar equation may be written for dz:

$$dz = \left(\frac{\partial z}{\partial y}\right)_x dy + \left(\frac{\partial z}{\partial x}\right)_y dx \qquad (A.16)$$

Substituting for dz from Equation (A.16) into Equation (A.15) gives

$$dx = \left(\frac{\partial x}{\partial z}\right)_y \left(\frac{\partial z}{\partial x}\right)_y dx + \left[\left(\frac{\partial x}{\partial y}\right)_z + \left(\frac{\partial x}{\partial z}\right)_y \left(\frac{\partial z}{\partial y}\right)_x\right] dy \qquad (A.17)$$

Equation (A.17) must be valid whichever two of the variables x, y, z are taken to be independent. If x and y are chosen to be independent, dy can be made zero and dx non-zero. This gives

$$\left(\frac{\partial x}{\partial z}\right)_y \left(\frac{\partial z}{\partial x}\right)_y = 1$$

or

$$\left(\frac{\partial x}{\partial z}\right)_y = 1 \bigg/ \left(\frac{\partial z}{\partial x}\right)_y \qquad (A.18)$$

Equation (A.18) is known as the reciprocal theorem.

If, instead, dx is made zero and dy non-zero, Equation (A.18) gives

$$\left(\frac{\partial x}{\partial y}\right)_z = - \left(\frac{\partial x}{\partial z}\right)_y \left(\frac{\partial z}{\partial y}\right)_x$$

or, using the reciprocal theorem,

$$\left(\frac{\partial x}{\partial y}\right)_z \left(\frac{\partial y}{\partial z}\right)_x \left(\frac{\partial z}{\partial x}\right)_y = -1 \qquad (A.19)$$

This result is known as the reciprocity theorem, and such equations are known as reciprocity relations.

For a fuller discussion of partial differentiation reference may be made to the following books:

Hilton, P. J. (1958). *Differential Calculus* (London: Routledge and Kegan Paul)

Hilton, P. J. (1960). *Partial Derivatives* (London: Routledge and Kegan Paul)

Sources for numerical values

American Institute of Physics (1963). *Handbook*, 2nd edn (New York: McGraw-Hill)

Hsieh, J. S. (1975). *Principles of Thermodynamics* (Tokyo: McGraw-Hill Kogakusha)

Kaye, G. W. C. and Laby, T. H. (1973). *Tables of Physical and Chemical Constants*, 14th edn (London: Longman)

Nordling, C. and Osterman, J. (1980). *Physics Handbook* (Bromley: Chartwell-Bratt)

Roberts, J. K. and Miller, A. R. (1951). *Heat and Thermodynamics*, 4th edn (London and Glasgow: Blackie)

Answers and hints to exercises

Chapter 2

1   (a) Isolated.
     (b) Closed.
     (c) Open.
     [See Section 2.3.]

2   (a) Non-equilibrium.
     (b) Non-equilibrium.
     (c) Equilibrium.
     (d) Non-equilibrium.
     [See Section 2.4.]

3   (a) Quasistatic.
     (b) Non-quasistatic.
     (c) Quasistatic.
     (d) Non-quasistatic.
     [See Section 2.2.]

4   (a) False.
     (b) False.
     (c) False.
     (d) False.
     (e) False.

Chapter 3

1   The isotherms do not intersect.
    [*Hint*: Apply the zeroth law to the hypothetical situation where the isotherms do intersect.]

2   Temperature is a property of a large assembly of atoms; the temperature of a single atom of such an assembly has no meaning.

3 Intensive coordinates: pressure, temperature, load. Extensive coordinates: volume, length.
[See Section 2.1.]

4 (a) False.
(b) False.

5 $\theta_R = 300.4$ units; $\theta_r = 680.7$ units.
[*Hint*: θ_R is defined by the equation

$$\theta_R = aR + b$$

where a and b are constants whose values are determined by the values assigned to the ice point and steam point. θ_r is similarly defined.]

Chapter 4

1 (a) The viscosity of the liquid.
(b) The diffusion of ions through the porous membrane.

2 $$U_f - U_i = W + Q$$
$$U_f - U_i = W + Q + |\Delta m| c^2$$

where $|\Delta m|$ is the magnitude of the change in mass resulting from fission and c is the speed of light in a vacuum.

3 (a) Reversible.
(b) Quasistatic.
(c) Non-equilibrium.
(d) Non-equilibrium.
(e) Quasistatic.
[See Sections 2.2 and 4.6.]

4 The heat absorbed by the system during process B is $Q_B = -70$ J. The minus sign indicates that the heat transfer is from the system.
[*Hint*: Because the complete process is cyclic, the total change in internal energy of the system ΔU is zero. Using an obvious notation,
$\Delta U = 0 = Q_A + W_A + Q_B + W_B$ and $Q_A = +100$ J, $W_A = -50$ J and $W_B = +20$ J.]

5 Since the interacting systems are otherwise isolated, $\Delta U_A + \Delta U_B = 0$.
Therefore, $Q_A + Q_B + W_A + W_B = 0$. But $W_A = W_B = 0$ and, therefore, $Q_A = -Q_B$.

6 $Q = 0$, $W = 0$ and, therefore, $\Delta U = 0$.

7

$$W = \int_{L_i}^{L_f} FdL = \int_{L_i}^{L_f} \frac{C}{L_0}(L - L_0)dL$$

$$= \frac{C(L_f - L_i)}{2L_0}(L_f + L_i - 2L_0)$$

Chapter 5

1

$$W = -\int_{V_i}^{V_f} pdV$$

(a) $pV = 100$

Therefore,

$$W = -\int_{1}^{25} \frac{100}{V} dV = -322 \text{ J}$$

The minus sign indicates that the work is done by the system.

(b) $p = 104 - 4V$

Therefore,

$$W = -\int_{1}^{25} (104 - 4V)dV = -1248 \text{ J}$$

Again the minus sign indicates that the work is done by the system.

2

$$W = -\int_{V_i}^{V_f} pdV$$

p is constant and equal to p_0;

$$V_i = V_0 \text{ and } V_f = nV$$

3   *Hint:* Imagine that the initially evacuated cylinder A is connected to another cylinder, B, shown dotted in Figure H.1, fitted with a frictionless, non-leaking piston. Let this cylinder B be of such a size that it contains just the amount of air that will enter cylinder A when the valve is opened. When the first small quantity of air enters cylinder A, the pressure in B falls below atmospheric pressure by a very small amount and the piston of cylinder B is pushed in under a constant pressure p_0.

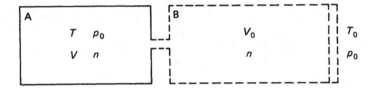

Figure H.1 *Notional model for the flow of air into an evacuated cylinder*

4   From Equation (5.6), if p is held constant,

$$dV = \left(\frac{\partial V}{\partial \theta}\right)_p d\theta$$

Now the cubic expansivity β is given by

$$\beta = \frac{1}{V}\left(\frac{\partial V}{\partial \theta}\right)_p$$

Therefore, at constant pressure,

$$\frac{dV}{V} = \beta d\theta$$

and, if β is constant,

$$V_f = V_i \exp\left(\beta(\theta_f - \theta_i)\right)$$

When β is small, $\exp\left(\beta(\theta_f - \theta_i)\right) \approx 1 + \beta(\theta_f - \theta_i)$ and

$$V_f \approx V_i + V_i\beta(\theta_f - \theta_i)$$

5

$$\text{Linear expansivity} = \frac{1}{L}\left(\frac{\partial L}{\partial \theta}\right)_F = -\frac{(L - L_0)}{2L\theta}$$

$$\text{Young's modulus} = \frac{L}{A}\left(\frac{\partial F}{\partial L}\right)_\theta = \frac{2LF}{A(L - L_0)}$$

6   Use the reciprocity relation

$$\left(\frac{\partial F}{\partial \theta}\right)_L = -\left(\frac{\partial L}{\partial \theta}\right)_F \left(\frac{\partial F}{\partial L}\right)_\theta$$

7   The work done on the surface is given by
$$W = \sigma(A_f - A_i)$$
where A_f and A_i are the final and initial surface areas, respectively. Since the bubble has an 'inside' and an 'outside' surface,
$$W = 2 \times 0.07 \times 4\pi\ (0.1^2 - 0.06^2)$$
$$= 1.13 \times 10^{-2}\ \text{J}$$

8   The work of magnetisation in an infinitesimal reversible process is given by
$$dW = V\mu_0 H_a dM \text{ (obtained from Equation 5.33)}$$
Using
$$\chi_m = \frac{M}{H_a} = \frac{C}{\theta}$$
and substituting for H_a gives the result.

Chapter 6

1   Atmospheric pressure = 76.3 cm of mercury. Assume that air obeys Boyle's law.

2   Apply Boyle's law. If p_0 is the initial pressure in the vessel of volume V, and v is the volume of the pump barrel, after the first stroke the pressure p_1 is given by
$$p_0 V = p_1(V + v),$$
and after n strokes the pressure p_n is given by
$$p_n = \frac{p_0 V^n}{(V + v)^n}$$

Therefore, after two strokes the pressure is 0.925 atm, a reduction of 0.075 atm, and 100 strokes are needed to reduce the pressure to 0.02 atm.

3   When $pV_m = f(\theta)$, the isothermal compressibility is equal to $1/p$:
$$\kappa_\theta = -\frac{1}{V_m}\left(\frac{\partial V}{\partial p}\right)_\theta \text{ and } \left(\frac{\partial V}{\partial p}\right)_\theta = -\frac{f(\theta)}{p^2}$$
When $pV_m = A + Bp$, the isothermal compressibility is given by
$$\kappa_\theta = \frac{1}{p} - \frac{B}{A + Bp}$$

4   230.3 J.
The work done on the gas is given by

$$W = - \int p\,dV$$

In an isothermal reversible process pV = constant, so that $pdV = -Vdp$ and

$$W = \int V\,dp$$

5   Let the volume of gas trapped at the pressure p_u be V. If the gas pressure is measured in mm Hg and h is measured in mm, when the reservoir is raised to give the mercury levels shown in Figure 6.6, the pressure of the trapped gas is $(p_u + h)$ mm Hg and, applying Boyle's law gives

$$P_u V = (p_u + h)\, Ah$$

where A is the area of cross-section of the tube. Then

$$P_u = \frac{Ah^2}{V - Ah}$$

If $Ah \ll V$

$$P_u = \frac{Ah^2}{V}$$

Chapter 7

1   Use the defining equations

$$\frac{Q_1}{Q_2} = - \frac{e^{\tau_1}}{e^{\tau_2}}; \frac{Q_3}{Q_4} = - \frac{e^{\tau_3}}{e^{\tau_4}}$$

$$\eta_1 = 1 + Q_2/Q_1 \; ; \; \eta_2 = 1 + Q_4/Q_3$$

2   The defining equation for τ is

$$\frac{Q_1}{Q_2} = - \frac{e^{\tau_1}}{e^{\tau_2}}$$

When $T = 100$ K,

$$\frac{Q_1}{Q_2} = - \frac{100}{273.16}$$

and, when $T = 273.16$ K, $\tau = 0$.

The temperature in thomsons corresponding to a temperature of 100 K is given by

$$-\frac{e^{\tau}}{e^0} = \frac{100}{273.16}$$

i.e. $\tau = -1.00$ Th.

$\left.\begin{array}{r} 3 \\ 4 \end{array}\right\}$ Use the equations

$$-W = Q_1 + Q_2 \text{ and}$$

$$\frac{Q_1}{Q_2} = -\frac{T_1}{T_2}$$

5 (b)

6 No. The engine is more efficient than a Carnot engine operating between the same reservoirs.

7 (c)

8 (d) The device is merely dissipating mechanical energy.

9

$$\text{C.o.P. (heat pump)} = -\frac{Q_1}{W} = \frac{T_1}{T_1 - T_2}$$

$$\text{C.o.P. (refrigerator)} = \frac{Q_2}{W} = \frac{T_2}{T_1 - T_2}$$

For the dependence of C.o.P on T_2/T_1 see Figure H.2.

10 See Figure H.3.

Chapter 8

1 Assume that the process is reversible.

$$\Delta S = \frac{Q}{T} = \frac{0.64 \text{ kWh}}{373\text{K}} = \frac{0.64 \times 10^3 \times 60 \times 60}{373} = 6180 \text{ J K}^{-1}$$

2 With an obvious notation

$$W + Q_1 + Q_2 + Q_3 = 0 \text{ (first law)}$$

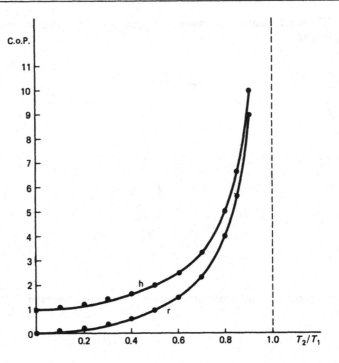

Figure H.2 *The dependence of C.o.P. on T_2/T_1. Curve h is for a heat pump and curve r for a refrigerator*

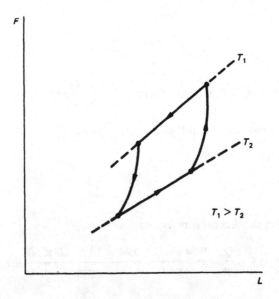

Figure H.3 *The Carnot cycle for a system consisting of a sample of rubber that obeys Hooke's law*

$$\frac{Q_1}{T_1} + \frac{Q_2}{T_2} + \frac{Q_3}{T_3} = 0 \text{ (second law)}$$

Remembering that the sign convention applies to the working substance,

$$W = -200 \text{ J}, \ T_1 = 400 \text{ K}, \ T_2 = 300 \text{ K}, \ T_3 = 200 \text{ K}, \ Q_1 = +1200 \text{ J}$$

Therefore,

$$Q_2 = -1200 \text{ J and } Q_3 = +200 \text{ J}$$

3   See Figure H.4.

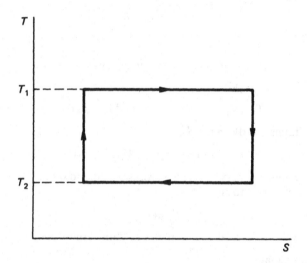

Figure H.4   *A Carnot cycle plotted on a graph of temperature* T *against entropy* S *is a rectangle for all systems*

The area enclosed by the curve is $-W$, where W is the work done on the working substance in one cycle.

4   The resistor does not undergo a change of state and, therefore, its entropy does not change. The entropy change of the water ΔS is given by

$$\Delta S = \frac{I^2 R t}{T}$$

$$= \frac{1 \times 25 \times 10}{280} = 0.9 \text{ JK}^{-1}$$

5   From the first law,

$$dU = -p dV + T dS$$

Therefore

$$\left(\frac{\partial U}{\partial V}\right)_S = -p \text{ and } \left(\frac{\partial U}{\partial S}\right)_V = T$$

6 The complete system is isolated from its surroundings and has a fixed volume

(a) Using an obvious notation,

$$dU_1 + dU_2 = 0$$
$$dV_1 + dV_2 = 0$$
$$dS_1 + dS_2 = 0$$

Therefore,

$$dU_1 = -dU_2; \ dS_1 = -dS_2$$

Using $dU = TdS - pdV$,

$$0 = T_1 \, dS_1 - p_1 \, dV_1 + T_2 \, dS_2 - p_2 \, dV_2$$

or, substituting for dV_2 and dS_2,

$$0 = dS_1(T_1 - T_2) - dV_1(p_1 - p_2)$$

Since, in an infinitesimal reversible process, neither dS_1 nor dV_1 is zero, for this identity to be satisfied,

$$T_1 = T_2 \text{ and } p_1 = p_2$$

That is, both mechanical and thermal equilibrium must obtain.

(b) In this situation

$$dS_1 = dS_2 = 0$$

Therefore, T_1 and T_2 can assume any values consistent with the condition for mechanical equilibrium: $p_1 = p_2$.

Chapter 9

1 Equation (7.18) leads to a defining equation for entropy of

$$dS = \frac{q}{e^\tau}$$

and an equation of state for 1 mol of molecules of ideal gas of

$$pV_m = Re^\tau$$

2 Let the number of moles of molecules initially in the vessels of volumes V_0 and V be n_1 and n_2, respectively. Then

$$p_0 V_0 = n_1 R T_0 \text{ and } p_0 V = n_2 R T_0$$

When the temperature of the vessel of volume V is changed to T, the numbers of moles of molecules change to, respectively, n_3 and n_4. Then

$$pV_0 = n_3 R T_0 \text{ and } pV = n_4 R T$$

Since $n_1 + n_2 = n_3 + n_4$, the result follows.

3   The cubic expansivity β is given by

$$\beta = \frac{1}{V}\left(\frac{\partial V}{\partial T}\right)_p$$

and for n moles of ideal gas $(\partial V/\partial T)_p = nR/p$.

4   For an ideal gas there is no change in internal energy when it undergoes a reversible isothermal change in volume. For such a process the first law becomes

$$TdS = q = -pdV = w$$

Since $pV = nRT$,

$$W = -\int_{V_i}^{V_f} pdV = nRT \ln\left(\frac{V_i}{V_f}\right)$$

and

$$Q = nRT \ln\left(\frac{V_f}{V_i}\right) = nRT \ln\left(\frac{p_i}{p_f}\right)$$

5   No. See Dalton's and Leduc's laws.

6   $p' = 1.003\, p$. Use the approach of Exercise 2.

Chapter 10

1   By definition

$$F = U - TS$$

For a closed hydrostatic system

$$dU = TdS - pdV$$

so that

$$dF = -pdV - SdT$$

At constant volume

$$\left(\frac{\partial F}{\partial T}\right)_V = -S$$

and, therefore,

$$U = F - T\left(\frac{\partial F}{\partial T}\right)_V$$

A similar proof follows for

$$G = F - V\left(\frac{\partial F}{\partial V}\right)_T$$

2

$$J = S - \frac{U}{T}$$

Therefore

$$dJ = dS - \frac{dU}{T} + \frac{U}{T^2}\,dT$$

Now, from the entropy form of the first law,

$$\frac{dU}{T} = dS - \frac{pdV}{T}$$

and the result follows.

3 Use the appropriate analogue of the Gibbs–Helmholtz equations, i.e.

$$U = F + T\left(\frac{\partial F}{\partial T}\right)_L \quad \text{and} \quad G = F - L\left(\frac{\partial F}{\partial L}\right)_T$$

together with the result that, in a reversible isothermal process, the work done on a system is equal to the increase in F. Then

$$\Delta U = \frac{L_0^2}{2}(2aT - 3bT^2)$$

$$\Delta G = \frac{5L_0^2}{2}(aT - bT^2)$$

Chapter 11

1 Writing $V = V(T, p)$, the change in volume in an infinitesimal process may be written

$$dV = \left(\frac{\partial V}{\partial T}\right)_p dT + \left(\frac{\partial V}{\partial p}\right)_T dp = V\beta dT - \kappa_T V dp$$

where β is the cubic expansivity and κ_T is the isothermal compressibility. At constant volume dV is zero and, assuming that κ_T and β are constant,

$$\beta \Delta T = \kappa_T \Delta p$$

When $\Delta T = 1$ K,

$$\Delta p = \frac{50 \times 10^{-6}}{14 \times 10^{-12}} = 3.6 \times 10^6 \text{Pa}$$

2 (a) Apply conservation of internal energy. The final equilibrium temperature T_f is given by

$$T_f = \frac{c_1 T_1 + c_2 T_2}{c_1 + c_2}$$

(b) Apply conservation of entropy. Then

$$T_f = T_1^{(c_1/c_1+c_2)} T_2^{(c_2/c_1+c_2)}$$

3

$$C_{p,m} = aT^3 + bT$$

$$dQ_R = C_p dT \; ; \; dS = \frac{dQ_R}{T} = \frac{C_p dT}{T}$$

Therefore,

$$\Delta S_m = \int_1^9 \left(\frac{aT^3}{T} + \frac{bT}{T} \right) dT$$

$$= 8.5 \times 10^{-3} \text{ J K}^{-1} \text{ mol}^{-1}$$

4 Write Newton's law of cooling in the form

$$\frac{dQ}{dt} = C_p \frac{dT}{dt} = - \text{constant} \, (T - T_s)$$

Integration gives

$$C_p \ln(T - T_s) = - K't + K''$$

where K' and K'' are constants.

The time for the temperature to fall from 333 K to 293 K is 8.2 min.

5 Use the method of cooling (Section 11.5.2). At 460 K the value of W is 25W and dT/dt is $- 0.154$ K s^{-1}, giving a specific heat capacity of 1080 J kg^{-1} K^{-1}.

6 Writing U_m as a function of T and p show that, at constant T,

$$\left(\frac{\partial U}{\partial p} \right)_T = - T \left(\frac{\partial V}{\partial T} \right)_p - p \left(\frac{\partial V}{\partial p} \right)_T$$

which, for the gas in question, is equal to

$$T\frac{dB}{dT}$$

7 Imagine that the air that enters the cylinder is contained in an imaginary piston, as in Figure H.1.

The work done by the atmosphere on the gas that enters the cylinder is $W = p_0V_0$, while the heat transfer is zero. The change in internal energy of the gas in an infinitesimal part of the process is given by

$$dU = C_V\,dT = -p\,dV$$

Therefore, in the complete process,

$$\Delta U = nC_{V.m}(T - T_0) = p_0V_0$$

But, $p_0V_0 = nRT_0$ (ideal gas equation) and $C_{p.m} - C_{V.m} = R$ (Equation 11.22). Therefore,

$$nC_{V.m}(T - T_0) = nRT_0 = n(C_{p.m} - C_{V.m})T_0$$

and

$$\frac{T}{T_0} - 1 = \frac{C_{p.m}}{C_{V.m}} - 1$$

$$\text{or } T = \gamma\, T_0.$$

8 Write $H = H(T,p)$ to give

$$dH = \left(\frac{\partial H}{\partial T}\right)_p dT + \left(\frac{\partial H}{\partial p}\right)_T dp$$

and use

$$dH = TdS + Vdp$$

to obtain expressions for the partial derivatives, modified, if necessary, with a Maxwell relation.

9 The approach outlined in Exercise 8 will enable all the equations to be derived.

10 Start with Equation (10.4):

$$dU - T_0\,dS < w$$

In this situation w may be written $w = -p_0dV$. Then

$$dU - T_0dS + p_0dV < 0$$

When the process is adiabatic, dS is zero and the equation becomes

$d(U + p_0V) < 0$ or $dH < 0$. The process will occur when $dH < 0$.

Chapter 12

1 The result is Figure H.5. In a reversible isothermal process T is a constant
   and so is U for an ideal gas. When an ideal gas changes its temperature
   reversibly, as in an isentropic process,

$$dU = C_V dT$$

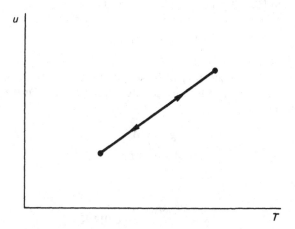

Figure H.5 *A Carnot cycle plotted on a graph of internal energy* U *against
temperature* T *when the working substance is an ideal gas with constant specific
heat capacities*

2 The fraction of the work recovered is 0.61.

3 The gravitational potential energy decrease as the ball-bearing falls a
   distance L is of magnitude MgL. The work done on the gas is

$$-\int_0^L (p - p_0)dV$$

where p is the pressure in the gas. Use $pV^\gamma = $ constant to obtain

$$p\gamma \frac{dV}{V} + dp = 0$$

Assume that the variations in p and V are very small. Then

$$dV = -A dy$$

where y is measured from the initial position of the bearing, and

$$dp = -\frac{p_0\gamma}{V} dV = \frac{p_0\gamma A}{V} dy$$

so that, using the condition that $p = p_0$ when $y = 0$,

$$p - p_0 = \frac{p_0 \gamma A y}{V}$$

Substituting for dV and $(p - p_0)$ gives the result.

4   Use Equation (12.29), which integrates to give

$$\Delta T = -\frac{T}{C_L} A E \lambda \Delta L$$

provided that ΔT is small. If c_L is the specific heat capacity under constant load, $C_L = A L \rho c_L$, where ρ is the density of the material. Further, if the applied mass is M,

$$\frac{Mg}{A} = E \frac{\Delta L}{L}$$

Therefore,

$$\Delta T = -\frac{T \lambda M g}{A \rho c_L} = -0.014 \text{ K}$$

5   Write $L = L(T, F)$ and apply the conditions that dF must be the same in both rods and that

$$dL_1 + dL_2 = 0$$

6   Use Equation (12.43), which, for an isentropic change, may be written

$$dT = \frac{T}{C_A} \frac{d\sigma}{dT} dA$$

When the change in temperature is small, this equation may be written

$$\Delta T = \frac{T}{C_A} \frac{d\sigma}{dT} \Delta A$$

The value of $d\sigma/dT$ is -0.24 mJ m^{-2} K^{-1}, the value of ΔA is -125.5×10^{-6} m^2 and the value of C_A is 17.5×10^{-6} J K^{-1}. Substitution gives

$$\Delta T = 0.56 \text{ K}$$

7   In the notation of Section 12.4

$$dS_T = \frac{dQ_T}{T} = V \left(\frac{\partial M}{\partial T} \right)_{B_a} dB_a$$

$$\chi_m = \frac{a}{T - T_c} \text{(Curie–Weiss law)}$$

$$= \frac{\mu_0 M}{B_a} \text{(definition)}$$

Therefore,

$$\left(\frac{\partial M}{\partial T}\right)_{B_a} = -\frac{B_a a}{\mu_0 T^2}$$

and

$$\Delta S_T = -\frac{Va}{2\mu_0 T^2}(B_{af}^2 - B_{ai}^2)$$

8   Use Equation (12.74):

$$\frac{dE}{dT} = 0.00034 \text{ V K}^{-1}$$

and

$$E_{350} = 0.063 \text{ V}$$

This gives a value for ΔH_T of 5.4 kJ mol^{-1}.

9   (a)

$$\eta(\text{theoretical}) = \frac{T_1 - T_2}{T_1} = \frac{825 - 275}{825}$$

$$= 0.67$$

(b)

$$\eta = -\frac{W}{Q_1} = -\frac{dW/dt}{dQ_1/dt}$$

Therefore,

$$0.3 = -\frac{(-1.2 \times 10^6)}{dQ_1/dt}$$

so that

$$\frac{dQ_1}{dt} = 4.0 \times 10^6 \text{ W}$$

$$= 4.0 \text{ MW}$$

(c)

$$\frac{dQ_1}{dt} + \frac{dQ_2}{dt} + \frac{dW}{dt} = 0$$

Therefore,

$$4 \times 10^6 + \frac{dQ_2}{dt} + (-1.2 \times 10^6) = 0$$

and

$$\frac{dQ_2}{dt} = -2.8 \text{ MW}$$

(d)

$$\frac{dm}{dt} c_p \Delta T = 2.8 \text{ MW}$$

$$\frac{dm}{dt} = \frac{2.8 \times 10^6}{4.2 \times 10^3 \times 5}$$

$$= 0.13 \times 10^3 \text{ kg s}^{-1}$$

Chapter 13

1   (a) Valid; (b) valid; (c) invalid. Apply Equation (13.13)

2                                    $V_c = 3b$

$$p_c = \frac{RT_c}{2b} - \frac{a'}{9T_c b^2}$$

$$T_c^2 = \frac{8a'}{27Rb}$$

Use Equations (13.4).
   If $\hat{T}$ is the reduced temperature and $\hat{V}_m$ is the reduced molar volume,

$$\left(p + \frac{3p_c}{\hat{T}\hat{V}_m^2} \right)\left(V - \frac{V_c}{3} \right) = RT$$

3   Use successive approximations for p, say, starting with $p = RT/V_m$

4

$$C_{p,m} - C_{V,m} = \frac{R}{1 - 2a(V_m - b)^2/RTV_m^3}$$

Use Equation (11.24).

5

$$\frac{T_B}{T_c} = \sqrt{6} = 2.45$$

Use the reduced equation of state and write as a series in $\hat{p}$. The Boyle temperature is that which makes the coefficient of the term in $\hat{p}$ equal to zero.

6

$$a'' = \frac{0.42748 \, R^2 T_c^{2.5}}{p_c} \; ; b'' = \frac{0.08664 \, R \, T_c}{p_c}$$

7 Start with the equation

$$C_{sat} = T\left(\frac{\partial S}{\partial T}\right)_{sat}$$

and combine with a suitable differentiation of

$$dU = TdS - pdV$$

8 Consider the isotherm ABKD of Figure 13.2. Let the volume of the system be V when it is in the state represented by the point K, let V_B be the volume corresponding to the state B, when all the substance is vapour, and V_D that corresponding to the state D, when all the substance is liquid.

Further, let c and $(1 - c)$ be the proportions by mass of the liquid phase and the vapour phase, respectively, in the state represented by the point K. Then,

$$V = c \, V_D + (1 - c) \, V_B$$

so that

$$c = \frac{V_B - V}{V_B - V_D}$$

Now, $V_B - V$ is proportional to KB on Figure 13.2 and $V_B - V_D$ is proportional to DB. Therefore,

$$c = \frac{KB}{DB} \text{ and } (1 - c) = \frac{KD}{DB}$$

so that

$$\frac{c}{1 - c} = \frac{m_l}{m_g} = \frac{KB}{KD}$$

Chapter 14

1 Use

$$W = - \int p \, dV$$

to show that the work done is about 7% of the enthalpy of vaporisation.

2 The triple temperature is determined from the intersection of the sublimation and vaporisation curves. $T_t = 195$ K. Equation (14.14) may be used to determine the enthalpy of vaporisation at the triple point. $H_{m.v} = 25.5$ kJ mol^{-1}.

3 Use Equation (14.19). c (saturated steam) $= -6.1$ kJ kg^{-1} K^{-1}.

4 29.2 kJ. Operate a Carnot refrigerator between the mass of water and a heat reservoir at a temperature of 20 °C.

5 Use Equation (14.12):

$$\frac{dp}{dT} = -13.6 \text{ MPa K}^{-1}$$

6

$$\frac{dp}{dT} = 3.6 \times 10^3 \text{ Pa K}^{-1}$$

$$= 3.55 \times 10^{-2} \text{ atm K}^{-1}$$

Therefore, $\Delta T = -18.3$ K, assuming that dp/dT is a constant and the boiling point of water is 354.9 K or 82 °C.

7 Start with the first law of thermodynamics and remember that both T and p are constant in a first-order phase change. Combine with $\Delta S = H_{12}/T$ and Equation (14.12).

8 Start with Equation (14.11) and use reciprocity and reciprocal relations.

Chapter 15

1 (a) Use Equation (12.44).
   (b) Use Equation (12.68).

2 The best refutation is probably that due to Einstein, who pointed out that thought experiments must at least be possible in principle. No real process

can be completely reversible nor can heat transfer be avoided altogether. In Nernst's cycle the slightest heat influx or irreversibility throws the system away from absolute zero.

3 Use Equation (M3) (page 113). Neither van der Waals' equation nor the ideal gas equation is in agreement with the result derived.

Chapter 16

1 Use Equation (16.2).
Joule coefficient $= -a/C_{V,m}V_m^2$

2 Examine the properties of Equation (16.19).

3 The result follows directly from Equation (16.11) and the definition of enthalpy. The first term in the square brackets indicates the departure from Joule's law and the second that from Boyle's law.

4 T_c is obtained using Equation (13.4). T_B is obtained by expressing van der Waals' equation in the form

$$pV_m = RT + \frac{RTb-a}{V_m} - \frac{RTb^2}{V_m^2} + \cdots$$

$T = T_B$ when $RTb - a = 0$. Then show that the equation of the inversion curve is

$$T_i = \frac{2a}{Rb}\left(1 - \frac{b}{V_m}\right)^2$$

The maximum value of the inversion temperature is then $T_{imax} = 2 T_B$.

5 Use Equation (16.19).

$$\Delta T = + 5.8 \text{ K}$$

6 The equation of the inversion curve is

$$\left(\frac{\partial \hat{V}_m}{\partial \hat{T}}\right)_{\hat{p}} = \frac{\hat{V}}{\hat{T}}$$

Determine $(\partial \hat{V}_m/\partial \hat{T})_p$ and substitute for this and for $\hat{T}$ in the above equation.

Chapter 17

1 Let 1 m³ of ideal gas contain n moles of molecules. Then

$$pV = nRT$$

If N is the number of molecules, $N = nN_A$ and R/N_A is Boltzmann's constant k. Therefore,

$$N = \frac{pV}{kT}$$

$$= 2.68 \times 10^{25} \text{ m}^{-3}$$

2

$$\text{Mean free path} = \frac{1}{N\pi\sigma^2}$$

where N is the mean number of molecules in 1 m^3. From Exercise 1, N is 2.68×10^{25} m^{-3}. Therefore, for nitrogen at s.t.p. the mean path is 3.96×10^{-7}m.

3

$$p = \frac{1}{3}\rho\overline{c^2}$$

The mass of 1 mol of nitrogen molecules is 28 g, which is 28×10^{-3} kg. Therefore,

$$\rho = \frac{28 \times 10^{-3}}{22.4 \times 10^{-3}} = 1.25 \text{ kg m}^{-3}$$

and $\overline{c^2}$ is 24.24×10^4 m^{2}s^{-2}, so that the root mean square speed is 492 ms^{-1}.

4 The mean free path (fast molecule approximation) is equal to $1/n\pi\sigma^2$, where n is the number of molecules in unit volume and σ is the (hard sphere) molecular diameter. The average separation of the molecules is $1/n^{1/3}$ and so the result follows.

Chapter 18

1 308.4 K.

2 For a layer of insulation having internal and external radii R and R_0, respectively, the total heat transfer coefficient $\overline{U}$ is given by

$$\frac{1}{\overline{U}} = \frac{1}{2\pi R_0 h} + \frac{\ln(R_0/R)}{2\pi k}$$

This has a minimum value obtained by putting

$$\frac{d\overline{U}}{dR_0} = 0$$

giving $R_0 = k/h$.

3

$$r_0 = \sqrt{r_1 r_2}.$$

4 The dispersion produced by a prism is non-uniform, so that the thermometer bulb intercepts a greater range of wavelengths at the red end of the spectrum, and in the near infra-red, than it does elsewhere.

5 Use $dU = TdS - pdV$ to show that

$$S = \frac{16\sigma T^3 V}{3c} + \text{constant}$$

Then

$$\left(\frac{\partial S}{\partial T}\right)_V = \frac{16\sigma T^2 V}{c}$$

and

$$C_V = \frac{16\sigma T^3 V}{c}$$

6

$$T^4 - T_W^4 = (T^2 - T_W^2)(T^2 + T_W^2)$$
$$T^2 - T_W^2 = (T - T_W)(T + T_W)$$
$$\text{when } T \approx T_W, \ T + T_W \approx 2T_W$$

Then Equation (18.36) becomes

$$\frac{dq}{dt} = 4A\alpha(T)\sigma T_W^3(T - T_W)$$

7 When $c_2/\lambda T \gtrsim 5$, $e^{c_2/\lambda T} \gg 1$.

Using this result and taking natural logarithms of both sides of the modified Equation (18.53) gives the result.

Bibliography and references

Bibliography

Adkins, C. J. (1983). *Equilibrium Thermodynamics*, third edition (Cambridge: Cambridge University Press)

Finn, C. B. P. (1986). *Thermal Physics* (London: Routledge and Kegan Paul)

Pippard, A. B. (1957). *The Elements of Classical Thermodynamics*, (Cambridge: Cambridge University Press)

Riedi, P. C. (1988). *Thermal Physics*, second edition (Oxford: Oxford University Press)

Zemansky, M. W. (1968). *Heat and Thermodynamics*, fifth edition (Tokyo: McGraw-Hill Kogakusha)

References

Amagat, E. H. (1870). *Comptes Rendus*, **71**, 67 [1]

Andrews, T. (1869). *Phil. Trans. Roy. Soc.*, **159**, 575

Andrews, T. (1876). *Phil. Trans. Roy. Soc.*, **167**, 421

Avogadro, A. (1811). *J. de Physique*, **73**, 58 [2]

Bernoulli, D. (1738). *Hydrodynamica* (Strasbourg: Dulsecker)

Boyle, R. (1662). *New Physico-Mechanical Experiments* (London) [3]

Dalton, J. (1802). *Memoirs of the Literary and Philosophical Society of Manchester*, **5**, 595 [4]

Dalton, J. (1808). *A New System of Chemical Philosophy*, Volume 1, Part 1 (London: Bickerstaff) [5]

Daniell, J. F. (1836), *Phil. Trans. Roy. Soc.*, **126**, 107

Désormes, C. B. and Clément, N. (1819). *J. de Physique*, **89**, 321, 428 [6]

Gay-Lussac, J. L. (1802). *Annales de Chimie*, **43**, 137

Gay-Lussac, J. L. (1809). *Memoires de Physique et de Chimie de la Societe d'Arcueil*, **2**, 207 [7]

Holborn, L. and Otto, J. (1926). *Z. Phys.*, **38**, 359 (and earlier papers)

Holborn, L. and Schultze, H. (1915). *Ann. Phys. Lpz.*, **47**, 1089

Hoxton, L. G. (1919). *Phys. Rev.*, **13**, 438

Joule, J. P. (1845). *Phil. Mag.*, Series 3, **26**, 369
Kamerlingh Onnes, H. (1901). *Leiden Comm.*, No. 71
Knudsen, M. H. C. (1915). *Ann. Phys. Lpz.*, **48**, 111   [8]
Ko, C. C. (1934). *J. Franklin Inst.*, **217**, 173
Langmuir, I. (1913). *Phys. Rev.*, **11**, 329
Lees, C. H. (1898). *Phil. Trans. Roy. Soc.*, **191**, 399
Marcus, P. M. and McFee, J. H. (1959). In *Recent Research in Molecular Beams*, edited by I. Esterman (New York: Academic Press)
Pitzer, K. S. (1941). *J. Amer. Chem. Soc.*, **63**, 2413
Rossini, F. D. and Fransden, M. (1932). *J. Res. Nat. Bur. Stand.*, **9**, 733
Scheel, H. and Heuse, W. (1912). *Ann. Phys. Lpz.*, **37**, 79
Simon, F. E. (1956). *Year Book of the Physical Society*, Vol. 1 (London: The Physical Society)
Swann, W. F. G. (1909). *Proc. Roy. Soc.*, **A82**, 147
Thomson, W. and Joule, J. P. (1853). *Phil. Trans. Roy. Soc.*, **143**, 357
Thomson, W. and Joule, J. P. (1854). *Phil. Trans. Roy. Soc.*, **144**, 321
Thomson, W. and Joule, J. P. (1860). *Phil. Trans. Roy. Soc.*, **150**, 325
Thomson, W. and Joule, J. P. (1862). *Phil. Trans. Roy. Soc.*, **152**, 579
Zartman, I. F. (1931). *Phys. Rev.*, **37**, 383

Notes

[1] Translations of some of Amagat's papers on the compressibility of gases are in *The Laws of Gases*, edited by C. Barus (New York: American Book Company, 1899)

[2] Translated in Alembic Club Reprint No. 4 (Edinburgh: Livingstone, 1961) and also in *The Origins and Growth of Physical Science*, Volume 2, edited by D. L. Hurd and J. J. Kipling (Penguin Books, 1964)

[3] Part II, Chapter V, 'A defence of the doctrine touching the spring and weight of the air' is in *The Laws of Gases*, edited by C. Barus (New York: American Book Company, 1899)

[4] Reproduced in *The Expansion of Gases by Heat*, edited by W. W. Randall (New York: American Book Company, 1902)

[5] Representative arguments from this book are in *The Origins and Growth of Physical Science*, Volume 2, edited by D. L. Hurd and J. J. Kipling (Penguin Books, 1964)

[6] N. Clément married the daughter of C. B. Désormes and, subsequently, often styled himself Clément-Désormes. This has sometimes led to confusion.

[7] A translation is given in *The Origins and Growth of Physical Science*, Volume 2, edited by D. L. Hurd and J. J. Kipling (Penguin Books, 1964)

[8] A summary of Knudsen's experimental work is given in his book *The Kinetic Theory of Gases: Some Modern Aspects* (London: Methuen, 1934)

Index

Printed in the United States
By Bookmasters